Dance on Screen

Dance on Screen

Genres and Media from Hollywood to Experimental Art

Sherril Dodds
Lecturer in Dance Studies
Department of Dance Studies
University of Surrey

791.43655
D64d

First published 2001 by
PALGRAVE
Houndmills, Basingstoke, Hampshire RG21 6XS and
175 Fifth Avenue, New York, N. Y. 10010
Companies and representatives throughout the world

PALGRAVE is the new global academic imprint of
St. Martin's Press LLC Scholarly and Reference Division and
Palgrave Publishers Ltd (formerly Macmillan Press Ltd).

ISBN 0–333–80145–8

This book is printed on paper suitable for recycling and made from fully managed and sustained forest sources.

A catalogue record for this book is available from the British Library.

Library of Congress Cataloging-in-Publication Data
Dodds, Sherril, 1967–
    Dance on screen : genres and media from Hollywood to experimental art / Sherril Dodds.
        p. cm.
    Includes bibliographical references and index.
    ISBN 0–333–80145–8
    1. Dance in motion pictures, television, etc. I. Title.
    GV1779 .D63 2001
    791.43'655—dc21
                                                        00–069217

10   9   8   7   6   5   4   3   2   1
10   09   08   07   06   05   04   03   02   01

Printed and bound in Great Britain by
Antony Rowe Ltd, Chippenham, Wiltshire

For Hazel and Dick Dodds

Contents

List of Plates

Preface

In writing this book I hope to offer the reader an opportunity to delve into the wealth and breadth of dance practices on screen. The material for the book came out of my doctoral research (Dodds, 1997) but my interest in the screen media has a much longer history. As an undergraduate student I was introduced to the art films of Peter Greenaway, in particular *Drowning by Numbers* (1988). The film is less concerned with action-driven narrative than with an opulent visual play. Not only do a series of quirky games litter the plot, but Greenaway also inscribes the film with puns, jokes and witty references. The screen is like a canvas for pastoral images in which corpses, fireworks and numbers are integrated into the landscape. What struck me was the primacy of formal structuring devices and compositional concerns.

During my MA studies, video became an essential tool for studying dance irrespective of its ephemeral nature or geographical locality. The easy accessibility of video allowed me to see Swedish choreographer Mats Ek's radical reworking of *Swan Lake* (1987) with its bald-headed, flat-footed swans. It was also during this period that I became aware of how the screen could take dance out of the theatre into alternative locations and new environments. Over and over again I watched the radical juxtaposition of people and settings in Pina Bausch's *The Lament of the Empress* (1989): a man carrying a wardrobe across a field; a bunny girl stumbling over ploughed land; and a woman sitting smoking on an armchair in the middle of a busy highway. Similarly, Lea Anderson's *Perfect Moment* (1992) introduced me to the versatility that the camera could bring to dance through constructing images that could only exist on screen. As a student I was intrigued by these multiple platforms for dance, especially with regard to how the screen opened up new choreographic possibilities.

It is not just dance that fascinates me. I am an avid spectator of all forms of screen media: reruns of *The Simpsons*, the latest Woody Allen film or just browsing around the World Wrestling Federation web site. It is this vocabulary of popular and art media images that informs how I look at dance on screen. Unfortunately, at present there is a lack of scholarly writing on the subject. Although a clutch of referee journal articles and features in popular magazines are in print, there is a dearth of reference texts that deal with the area of screen dance. The most recent is *Parallel Lines: Media Representations of Dance* (Jordan and Allen, 1993). This anthology provides a valuable collection of articles on a diverse range of subjects; however, since its publication there has been a proliferation of dance works made specifically for the screen, an area which is only touched upon in *Parallel Lines*. I sense this situation might change, however. As a field of academic study, dance on screen is rapidly evolving. I am increasingly coming into contact with other scholars interested in this area and there are numerous students keen either to write dissertations on the subject or else to create their own dance films. Promising developments are taking place in higher education. A number of departments now run modules or courses in 'dance in the media' or 'dance technology' and this type of institutional recognition firmly places screen dance within a scholarly framework. This comment is not intended to undermine or disregard the artistic context out of which screen dance emerges; rather it is a call for a critical dialogue that can allow practitioners, students and scholars to analyse, interpret and evaluate this rich field of dance practice.

It therefore appears to be a prime time for a text devoted to the subject of dance on screen. This book sets out to provide a comprehensive introduction to the diversity of screen dance forms through cultural, economic, critical, artistic, historical, technical and theoretical perspectives. Although the book primarily addresses images of dance on film and television, in acknowledgement of the related and rapidly expanding area of digital screen dance, developments in this field form a small part of the discussion. Chapter 1 commences with an overview of the cultural context in which contemporary screen dance is situated, and considers some of the different rationales and approaches to putting dance on screen. It traces the historical developments of the field, from the early flickering dance films at the beginning of the century, through the advent of television, to the recent experimentation between dance and digital media. This lineage is followed by an examination of the current artistic context in which screen dance is located, contrasting the writings of dance critics with the views of screen dance practitioners. To bring this chapter to a close, a technical examination of the 'screen' is undertaken to provide the groundwork for the case studies that follow. Chapter 2 offers some rich examples of the field in practice in order to address the diverse ways in which the screen media have dealt with the dancing body. The chapter begins at the commercial end of screen dance by looking at 1980s Hollywood dance films, television advertising and music video. From this consumerist imagery the remainder of the chapter turns to art dance on television. It deals both with television adaptations of stage dance and early examples of dance made for the camera.

Whereas the first two chapters provide a general introduction to dance on screen, the remaining chapters focus specifically on dance that is conceived and choreographed for television. Chapter 3 presents the concept of 'video dance', an experimental area of work that explores the creative interface between dance and television. It is suggested that video dance, as a relatively new genre of dance on screen, can be characterized by cutting-edge imagery and innovative filming techniques. Drawing on examples from the BBC's *Dance for the Camera* series and Channel 4's *Tights Camera Action!*, Chapters 3 and 4 investigate the distinctive character of video dance. Chapter 3 is concerned with how the televisual apparatus acts on the dancing body and the implications of this for choreographic practice, the performing body, positions of spectatorship and critical readings. Chapter 4 then addresses the impact of dance practice on television. In this chapter it is argued that the choreographers of video dance are rooted in a post-modern stage dance tradition. Drawing on literature from the disciplines of dance studies and film/television studies, Chapter 4 examines how video dance employs postmodern stage dance strategies that challenge notions of realism, linear narrative and psychologically motivated characters, and call into question the way in which the television screen is perceived.

Drawing on the contextual and analytic work of Chapters 1 and 2 and the discursive findings of Chapters 3 and 4, Chapter 5 sets out to construct a theoretical paradigm with which to conceptualize video dance based on the notions of 'hybridity' and 'fluidity'. In addition to dance and television, it is argued, the body intersects with other aesthetic practices and intellectual frameworks. First, Chapter 5 addresses the notion of a 'consumer body'. It considers the similarities between video dance and the imagery of television advertisements and music videos and the extent that video dance merges and overlaps with these promotional texts. Second, Chapter 5 examines the 'technological body'. Drawing on Benjamin's (1973) thesis of 'mechanical reproduction' and more recent writings on 'new digital technologies', it investigates the technological implications of video dance in terms of how the body is conceived. Chapter 5 concludes with a discussion of how video dance disrupts existing boundaries of dance and television to become a medium of artistic challenge and creative vitality.

I hope that this book appeals not only to the dance community but also to any scholars, students and practitioners with an interest in media studies, visual culture, film and television, digital technologies, the performing arts and the body as a locus of social meaning. Finally I would like to thank Palgrave for continuing to support the field of dance scholarship.

SD

Acknowledgements

I would like to thank my colleagues at the Department of Dance Studies, University of Surrey for their continual support and encouragement, and Chris Jones from the National Resource Centre for Dance for her precious advice on the initial book proposal and helpful hints en route. Thanks are also due to the professionals from the field of screen dance who have contributed in multiple ways to the writing of this book: Bob Lockyer from the BBC; Rodney Wilson from the Arts Council of England; Anne Beresford from MJW Productions; Katrina McPherson who was meticulous in her explanations of technical terminology; and the practitioners, Tom Cairns, Anna Pons Carrera, Emma Gladstone, David Hinton, Wendy Houstoun, Alison Murray and Margaret Williams, who kindly gave up time from their hectic schedules to be interviewed. I also thank the Arts Council of England and BBC Television, and Margaret Williams for allowing the use of their photographs in this book. I would also like to thank my parents Hazel and Dick Dodds, and my partner James Powell who never failed to be a cheery companion as I agonized over different stages of the writing process. Finally, my sincerest thanks go to Dr Theresa Buckland for having faith in this particular project and for being such a generous mentor and friend.

1
Dance on Screen: a Contextual Framework

Visual culture in the late twentieth century

As we face the dawn of a new millennium and cast a fleeting glance back at the late twentieth century, it is the mass media and its debris of popular artefacts that has come to characterize our everyday lives. We inhabit a culture of mass production in that media technologies can facilitate the rapid recording, reproduction and multiplication of sounds, texts and images. These apparati of communication comprise radio (and related systems of sound reproduction), film, print media, television and, more recently, digital formats such as CD-Roms and the internet (Kellner, 1995). The products of the mass media are constructed to appeal to a wide audience in order to generate vast financial profit. Hence the notion of 'mass' pertains to both production and reception. It is apparent that media culture is saturated by visual signs; two-dimensional, technologically produced representations. The images of film, television, videos, magazines, advertisements, computer graphics and electronic games bombard both public and private space. Within this visual landscape images have become the common currency. As everyday spectators of this distinctive cultural existence, we are adept at reading the complex relay of intertextual images. To sum up the prevalence of two-dimensional signs, the late twentieth society has been described as the 'society of spectacle' (Jameson, 1991). It is therefore not surprising that dance is embedded within the visual fabric of media technologies.

Since the inception of film and television, the former at the turn of the century and the latter in the early 1930s, dance has had a close involvement with the two media (de Marigny, 1988). The relationship is a reciprocal one: screen images intercept live dance performance and dance is translated to, or designed for, the screen. The projection of film and television images within live performance is no recent phenomenon. During the early 1900s, Georges Méliès created several films for use in stage productions (Pritchard, 1995–96). In 1966 Trisha Brown performed *A String* with a film projector strapped on her back which reflected her image onto the walls and audience, and in *Twice* (1970) Hans van Manen transmitted a pre-recorded version of the work on an overhead screen simultaneously with the live performance (Schmidt, 1991).[1] The use of film and video has continued with more current performances. British choreographers Lloyd Newson, Mark Murphy and Rosemary Butcher, and their French counterparts Jean-Claude Gallotta, Daniel Larrieu, Joelle Bouvier and Regis Obadia (Bozzini, 1991), have all employed film or video projections within recent works. The blurring of boundaries between live and virtual performance is perhaps the latest succession in this evolution with the use of digital images in stage dance (Hansen, 1998).

Aside from actually screening film, video and even computer-generated images in performance, representations of media technologies slip into stage dance through less direct routes. Some dance practitioners have drawn on specific films as a starting point for choreographic ideas. Ian Spink's *Further and Further into Night* (1984) literally 'quotes' movement sequences from the Alfred Hitchcock film, *Notorious* (1946), and Lea Anderson's *The Bends* (1994) alludes to the Marx Brothers' *Monkey Business* (1931), the German war epic *Das Boot* (1981) and the art film *Performance* (1970). Similarly, Jean-Claude Gallotta is said to be influenced by film maker Jean-Luc Godard and the stage works of Philippe Decouflé and Régine Chopinot are likened to advertising and video clips (Bozzini, 1991). While some artists deal with the content of screen media, others look to form. Pina Bausch's live work employs filmic devices that manipulate speed, such as slow motion or the acceleration of events (Müller, 1984), and montage structures that are indebted to the fragmented character of editing (Rosiny, 1990a). Meanwhile the two-dimensional look that typifies Lea Anderson's choreography owes much to her references to cinematic and televisual arts (Dodds, 1995–6).

While screen images have entered live performance in a variety of ways, dance can also be seen on film and television in a multiplicity of genres. For instance, in Hollywood musicals and the popular dance films of the 1970s and 1980s such as *Saturday Night Fever* (1977), dance forms a major component of the work. Dance can also provide a context to explore pertinent social issues: *The Red Shoes* (1948) takes the torturous nature of 'artistic genius' as its central theme and *They Shoot Horses Don't They* (1969) deals with the marathon dance competitions of the Depression. Elsewhere there are small, but nevertheless memorable, snippets of dance in mainstream Hollywood films. Al Pacino is noted for 'dancing a mean impromptu tango' (Grant, 1995, p. 643) in *Scent of a Woman* (1992) and John Travolta and Uma Thurman execute a sultry partner dance in *Pulp Fiction* (1994). Similarly, dance is screened on television in a diversity of contexts that include advertisements, pop music videos, 'light entertainment' shows, fly-on-the-wall documentaries and arts magazine programmes. Art dance on television ranges from live or recorded versions of existing stage work through to dance that is conceived and choreographed specifically for the camera. It appears that dance functions on television for a variety of different or overlapping reasons; these include popular entertainment, artistic innovation, commercial gain, the construction, documentation and preservation of the dance canon and so on.

What we actually see on screen is shaped by a number of factors. Film and television production is both labour and capital intensive and therefore much programming is dependent on economic decisions (Monaco, 1981). For instance, full-length ballets tend to be recorded during live performances due to the large scale of the work and the number of personnel involved; this, however, places restrictions on possible camera positions and may limit each shot to one 'take'.[2] Film is a particularly expensive medium and consequently it is often those with financial power who commission or produce work. Although video is relatively cheaper, practitioners still need access to time and equipment. And even if independent video (or film) artists succeed in creating their own work, there is no guarantee that it will ever be screened outside the arena of small-scale festivals. Production and broadcasting policy are undoubtedly determined by political components (Brooks, 1993). Funders, executive producers and commissioning editors largely control the films and programmes that are made and these decisions constantly negotiate the tensions between artistic innovation and financial reward (Tegeder, 1985). While a few television and film companies make a rare attempt to privilege aesthetic concerns, the wider cultural context is one that favours high audience figures and commercial gain. Although the spectator is empowered in relation to whether or not she or he chooses to watch a programme or film (Brooks, 1993), the work shown on the television network and in public cinemas is subject to broadcasting policy and the discourses of censorship.

In addition to economic and political issues, technical and aesthetic constituents affect the images that we see on screen. For instance, certain technical criteria contribute to the type of images that can be made: the quality of film celluloid versus video tape; the big screen as opposed to the small screen; the viewing conditions of the movie theatre in contrast to the domestic space of the home; and the technological possibilities of the mechanical film apparatus in relation to the electronic (or digital) video camera. Closely tied to such technical factors are the aesthetic conventions that have come to characterize film and television. For instance, the close-up has become a dramatic device for relaying psychological perspectives through the actor's facial expression, and parallel action scenes result from film and television's ability to cut between two series of separate events. As Monaco (1981) notes, the codes and conventions of an art form are inextricably linked to the technical features of the medium. Consequently, the dance that is screened on film and television is influenced by all of these economic, political, technical and aesthetic components and, in turn, the images that are screened shape the spectator's perceptions of dance.

In order to pursue images of dance on screen in greater depth, the remainder of Chapter 1 sets out to present a contextual overview. It considers the historical developments of screen dance and then focuses on its current artistic context. The chapter examines how screen dance is located within critical writings and how this contrasts with the views of the practitioners who create dance for the camera. To bring the chapter to a close, a comparative study is made between the live body and the screen body, the theatre setting and the television context, in order to provide a technical framework. This forms the starting-point for the case studies that follow in Chapter 2.

Histories of dance on screen

From the initial developments of film in 1895, it became apparent that dance was particularly compatible with the filmic form: both film and dance are characterized by motion and the art of editing shares similarities with the rhythmic component of dance (Pritchard, 1995–96). This mutual compatibility is evident in that some of the first cinematic examples feature moving objects or people as their subject matter (Delamater, 1981). In addition to images of speeding trains and factory employees leaving work, early film makers turned to the music hall as a source of inspiration. The first commercial screening in the United States on 23 April 1896 included two young women performing a 'parasol dance' and the popular entertainer, 'Annabelle the Dancer' (Delamater, 1981). Film maker Louis Lumière recorded a wide variety of indigenous dance forms (Pritchard, 1995–96), and Fernand Léger's *Ballet Mécanique* (1924–25) is a film exploration of rhythmic structures in movement and music (Monaco, 1981). The attraction of film also extended to dancers themselves. In 1919, Loïe Fuller made a 35mm colour film, *Le Lys de la vie* (Bozzini, 1991), which focuses on the manipulation of light and movement.

The new medium of film was an opportunity to document dance in both theatrical contexts and social settings. Anna Pavlova's performance of *The Dying Swan* (1924) is captured for posterity on celluloid and there are many film excerpts of the Charleston craze of the late 1920s. Georges Méliès, another key figure in the early developments of cinema, employed dancers to perform in his 'fantasy films' (Bozzini, 1991). Although Méliès originates from a theatrical background, within his films, 'the possibilities extended far beyond the stage and he was soon experimenting with multiple exposure and stop-action filming, editing his works in the camera as he made them' (Pritchard, 1995–96, p. 30). Indeed many stars of the silent era either came from dance backgrounds or studied dance to enhance their movement quality on screen: Rudolph Valentino began his career as a dancer in vaudeville and cabaret, and Charlie Chaplin is often credited for his dancerly physicality (Delamater, 1981).

At the end of the 'silent era'[3] in 1928, a whole genre of dance film emerged (de Marigny, 1988). The 1930s can be marked as the beginning of Hollywood's domination of the film industry, and with this came the creation of the 'musical'. From the 1930s to the beginning of the 1950s there was a proliferation of film musicals produced by major Hollywood studios such as MGM, RKO and Warner Brothers (Monaco, 1981). Such films can be characterized by their elaborate song and dance routines, performed by star names such as Fred Astaire, Ginger Rogers, Gene Kelly, Eleanor Powell, Cyd Charisse and Ruby Keeler (Stoop, 1984). The dance content of film musicals tends to be rooted in popular forms such as tap, vaudeville and social dance styles, rather than in theatre art dance (Barnes, 1985). In the Hollywood musicals, the dance and music are the primary components while the narrative takes a secondary position (Allen, 1993).

One of the most interesting aspects of the musical genre is that dance routines were choreographed specifically for the film medium, and Busby Berkeley was one director who filmed dance in an innovative way. Berkeley's dance routines draw on a tradition of Broadway revue in that they feature grand stairways, revolving platforms, thrust walkways and choruses of dancers, and tend to privilege visual experience over physical expression (Delamater, 1981). Berkeley is known for choreographing complex geometric dance routines that were filmed in such a way so as to create abstract, mobile patterns (Monaco, 1981). He manipulated scores of female dancers in precise, military motifs and the movement content generally depended on a series of poses rather than any established dance techniques. In Berkeley's work the camera is very much a participant in the dance. For instance, he often used tracking shots[4] to move along the lines of women and sometimes employed close-ups to show off each woman's face.[5] His signature mark, however, is the 'top shot'. He regularly placed the camera overhead so that his circular designs could be viewed as elaborate, kaleidoscopic effects (Delamater, 1981).

Another key individual who exerted a considerable influence on the relationship between dance and film is dancer-choreographer Fred Astaire. Unlike Berkeley, Astaire came from a musical comedy stage tradition and worked with solos and duets rather than choruses (Delamater, 1981). His famous dance partnership with Ginger Rogers formed the basis of numerous Hollywood musicals that include *The Gay Divorcee* (1934), *Top Hat* (1935) and *Shall We Dance?* (1937). Astaire drew on an eclectic movement vocabulary that includes ballroom, tap, jazz and, at times, a trace of ballet. Whereas Berkeley's dance extravaganzas marked a suspension of the plot, Astaire integrated the songs and dance into the narrative as far as possible (Delamater, 1981). Reputedly a perfectionist, Astaire developed some specific ideas on the way in which dance should be choreographed and filmed. Mueller (1984) states, 'the idea was to put the [film] medium at the service of the dance' (p. 132). Astaire constructed full-length dance numbers, unlike some choreographers who would simply edit together a series of short phrases. The dance was of paramount importance to Astaire and for this reason he insisted on full body shots, the dancers being captured in a 'tight frame', and a limited number of cuts. He rarely employed special effects or 'arty perspectives', and used 'reaction shots'[6] sparingly. The result is that the dance is seen as clearly as possible without being distorted through the filmic apparatus (Mueller, 1984).

The demise of the Hollywood system, and its hold over the film industry, took place in the early 1950s; yet dance continued to be seen in a variety of film contexts. The Hollywood musicals of the 1930s and 1940s were followed by film adaptations of stage musicals such as *Oklahoma!* (1955) and *Cabaret* (1972). As mentioned earlier, a number of narrative films use dance as an intrinsic feature of the plot, as with the critically acclaimed *The Red Shoes* (1948), or else employ dance as a vehicle for other concerns. An example of the latter is *Nutcracker* (1982), a soft pornography film set in the context of a ballet company (Allen, 1993). It is only at the end of the 1970s and throughout the 1980s that a new genre of dance film emerged. This era marks a proliferation of dance films, such as *Saturday Night Fever* (1977), *Dirty Dancing* (1987), *Fame* (1980), *Flashdance* (1983), *Staying Alive* (1983) and *Footloose* (1984), which tie in with the general health and fitness boom that flourished during this period (Buckland, 1993). Although popular music often forms the soundtrack of these films, they are not 'musicals' as identified above; rather, they are classic narrative films that use dance as a metaphor for social identity, romantic fulfilment and other 'fantasies of achievement' (McRobbie, 1990; Buckland, 1993).

With the exception of those made in the early years, the films discussed so far have all been produced within a mainstream Hollywood context. It is, however, worth making some note of the relevance of avant-garde cinema to screen dance. Several practitioners who have worked within experimental film-making practices were initially involved in dance. Wendy Toye, Yvonne Rainer and Sally Potter are three such individuals whose experience of dance informs their cinematic work (Pritchard, 1995–96). One of the most significant film makers, located within the avant-garde, in relation to dance on film is Maya Deren (Maletic, 1987–88). As a woman film maker, Deren was something of a rarity in the 1940s, eschewing Hollywood in favour of independent production and distribution (Clark et al., 1984). Deren is cited as being an innovator of 'chore-cinema', an art form in which the dance and the camera are inextricably linked. In 1945 Deren made *Study in Choreography for the Camera*, which she describes as 'a dance so related to camera and cutting that it cannot be performed as a unit anywhere but in this particular film' (cited in Satin, 1991, p. 41).

Study in Choreography for the Camera is an exploration of time and space through the formal apparatus of film and dance. In the film, Talley Beatty traverses diverse locations in the space of a few seconds; for instance, he commences a phrase of movement in a forest but then continues it in a living-room and then a courtyard (Clark et al., 1988). The continuity of the dance sequence remains, while the environment changes radically. Although this is Deren's only film that uses an explicit element of dance, her work is often described in terms of a dance sensibility. Through the manipulation of the camera, there is a strong choreographic quality to her films which is also enhanced by the movement on the screen and the style of edit (Satin, 1991). In *Ritual in Transfigured Time* (1946) two women sit opposite each other winding wool. One woman is shown in 'real time' while the other is shot in slow motion, which gives a fragmented feel to the action. Similarly, a party scene occurs during which the absence of any sound calls attention to the dancerly quality of the gestures and poses. The play on movement is further enhanced with the use of a freeze-frame device so that the 'stop-start' pedestrian action becomes formalized and episodic.

In much the same way that dance features in film from the early developments of the medium, dance was also regularly broadcast during the initial years of television.[7] Between 1932 and 1935 the BBC set up an unofficial broadcasting service. Although there would have been an element of popular and social dance styles within television variety shows, Penman (1986, 1993) documents a predominance of ballet during this period, which would have appealed to the sensibilities of the middle-class viewers who made up the television audience through these formative years. The policy to broadcast ballet was also due to the particular role of the BBC at the time. Eisele (1990) states that 'the emergence of ballet happened for the first time only in Europe, where state or publicly financed and controlled broadcasting systems were committed to a cultural mandate and not commercially oriented as in the USA' (p. 15).

In 1936 the BBC television service was officially launched and for the next three years ballet continued to be shown. Marie Rambert's Mercury Ballet was the first company to appear on the new service and, in addition to other choreographers, Anthony Tudor created eight new ballets specifically for television (Penman, 1993). Between 1936 and 1939 40 existing ballets were also screened, which gives some indication of the proliferation of dance on television in these early years. It is also of note that the BBC did not focus exclusively on ballet, but occasionally broadcast other dance forms such as Margaret Morris's 'free-form' dancing, Uday Shankar's Indian Dance Company, and the Spanish dancer, Argentinita (Rowson Davis, 1982–83). The service, however, abruptly closed down in 1939 with the beginning of the Second World War.

Initially, both recording and viewing conditions were primitive. The studio spaces were small and, once they were filled with the bulky television equipment, this left little room for the dancers. These limitations led to a predominance of solo or duet work, with little room for travelling steps. Due to the poor-quality television image during this era the dancers were also required to modify their costumes to very basic designs and wear layers of thick stage make-up (Rowson Davis, 1982–83). In 1946, following the end of the war, the BBC began transmitting once again and its dance programming continued to be just as prolific. Between 1946 and 1954 the BBC also began to screen a number of seasons by foreign companies and set about picking up new audiences with such programmes as *Ballet for Beginners* (Penman, 1993).

In 1954 Margaret Dale was appointed as a BBC producer. This was a significant choice as she was an ex-dancer from the Sadler's Wells Ballet. Towards the end of the 1950s, with much improved technical facilities, Dale began to broadcast large-scale works from the classical repertoire for television. Dale was also responsible for introducing Russian ballet to the British public through screening performances by the Bolshoi and Kirov Ballets. During this period another young producer, Patricia Foy, presented dance through other outlets such as gala performances and music programmes. In addition to performance-based programmes the late 1950s saw the rise of arts documentaries (Penman, 1987, 1993; de Marigny, 1988). The first such programme, *Monitor*, broadcast in 1958, set a precedent for other 'arts series' such as *Aquarius*, *The South Bank Show* and *Omnibus*.

During the 1960s television technology developed so that outside broadcasts could be made and the BBC began to purchase programmes from other television companies (Penman, 1987). This policy extended the coverage of dance and therefore allowed the public to see stage productions that they might otherwise not be able to see, whether for financial or geographical reasons (Brooks, 1993). Another purpose of recording stage dance for television was in order to ensure the conservation of existing works. By this period certain individuals were beginning to explore various methods for the recording of theatre dance. Swedish choreographer and director Birgit Cullberg initially propounded that a single camera should be held in a fixed shot without any movement or cutting. Although she retained the notion that the dancers should move while the camera remains static, she eventually began to favour a greater number of cuts as it gave the film a clearer sense of pace and dynamic (Eisele, 1990; Maletic, 1987–88). Other choreographers were meanwhile concerned with making work specifically for television. Hans van Manen's *Kaïn en Abel* (1961) is a television ballet shot on location. It includes a thriller-like chase through the streets and a *pas de deux* on a moving barge.

One figure who played an influential role in the developments of dance on video is American choreographer Merce Cunningham. He began investigating this area in 1974 in collaboration with video/film maker Charles Atlas. Cunningham comments that the appeal of work-ing with video is that the results can be seen immediately and it pro-vides another way of looking at dance (Grossman, 1979). Cunningham's first collaboration with Atlas was *Westbeth* (1974). The piece is an exploration of how video can affect time, space and energy and each section of the work deals with a particular 'video dance' problem. For instance, one section is concerned with the intimacy of the camera. Cunningham noted that, if the dancers stay too far from the screen, the spectator loses contact with them. In response, Cunningham and Atlas choreographed the movement and the camera in such a way that the dancers regularly came in close proximity to the camera. Another difficulty that Cunningham recognized was the triangular camera space, which is exceedingly narrow at the front and then tapers out-wards. Therefore in *Fractions* (1978, directed by Charles Atlas) Cunningham groups the dancers in a wedge-like formation with one dancer close to the camera and lines of two and three dancers behind (Maletic, 1987–88).

The other director with whom Cunningham has closely collaborated is Elliot Caplan. Caplan was employed by the Cunningham Dance Foundation in 1977 and is currently responsible for all of the com-pany's work in film and television (Rosiny, 1990b). For *Changing Steps* (TV 1989), Caplan designed long and complex camera moves to enable the dancers to execute full sections of movement material. In this work he allowed the dancers to 'break the frame' so that the camera would travel past them as they continued to move. In contrast, with *Points in Space* (1986) Caplan purposely kept all of the dancers within the frame (Rosiny, 1990b). One of the innovations that Cunningham has brought to television dance is that he utilizes the whole rectangle of the screen; he regularly places dancers at the edge of the frame, thus decentraliz-ing the screen space (Bozzini, 1991). Whereas *Points in Space* was tightly scripted, *Changing Steps* was largely structured in the editing suite. These different approaches highlight Cunningham's willingness to experiment with the medium. Indeed, not only is Cunningham's stage dance adapted for the screen, but his screen work has also been translated to stage. Both *Fractions* and *Channels/Inserts* (1982) were origi-nally made for the screen, but then reworked for the stage. Significantly, his work in video has clearly influenced his stage choreography. Cunningham comments that the precision of movement is given greater clarity on screen, and therefore he now gives more concentra-tion to detail in his stage dance (Percival, 1980).

While Cunningham was exploring dance and video in the United States during the 1970s, in the United Kingdom Bob Lockyer (currently the BBC Executive Producer for Music and Dance) began directing contemporary dance pieces for television. Lockyer developed a long involvement with London Contemporary Dance Theatre (LCDT) and was responsible for screening several LCDT works choreographed by Robert Cohan (de Marigny, 1988; Lockyer, 1993). In 1985 Cohan was commissioned by the BBC to create a dance work specifically for television and the piece, entitled *A Mass for Man*, was again directed by Lockyer. To some extent the work explores a number of televisual possibilities: it opens with close-up shots of feet and legs and images of destruction and third-world famine are later interspersed with shots of the dance. Yet the movement is largely choreographed with stage conventions in mind: the slow pace of the movement is somewhat laboured and the continuous use of full body shots makes for predictable viewing.

Throughout the 1980s two dance seasons were screened on BBC television, *Dance Month* (1980) and *Dance International* (1983), both featuring works from the classical repertoire or from mainstream contemporary dance companies. In contrast, Channel Four was able to screen more innovative and experimental dance (Allen, 1993; de Marigny, 1988), since it was not bound to mainstream appeal. Penman (1993) states that Channel Four's 'remit to cater for minority interests liberated it, to an extent, from the ratings war' (p. 119). The Commissioning Arts Editor at the time, Michael Kustow, was responsible for developing this cutting edge programming policy for dance. In the mid 1980s Kustow commissioned a series titled *Dance on 4* (1983) which was to become a cornerstone for innovative dance on television. *Dance on 4* featured stage works from the more experimental end of dance, such as Tom Jobe's *Run Like Thunder* (TV 1984) and Pina Bausch's *1980* (TV 1984) (de Marigny, 1993; Rubidge, 1993).

With vastly improved studio conditions and more sophisticated recording equipment, series such as *Dance International* and *Dance on 4* are a far cry from the early days of television. These programmes played an essential part in bringing dance to a television audience; yet they also highlighted some fundamental difficulties in terms of adapting or reworking dance for television.[8] It became evident that, although these works were being reconceived and filmed for television, they were often presented with a stage perspective in mind (Rubidge, 1993). At this time choreographers were unaware of the possibilities that the television medium could bring to dance, while directors were equally unfamiliar with the codes and conventions that characterized dance. It was for this reason that the Channel Four *Dancelines* (1987) project came about.

Dancelines was set up as a means to address some of the aforementioned problems with regard to adapting and creating dance for television. As well as a number of dancers and technical personnel, the collaboration involved director Terry Braun, lighting designer Peter Mumford, and choreographers Siobhan Davies and Ian Spink. Rubidge (1993) states: 'The intention was that each group of artists should learn about the intricacies of the other's medium, and about differing artistic approaches within each medium, whilst addressing the problems that their nascent ideas for works for television raised' (p. 189).

A number of criticisms levelled at the programme accused it of using too much technical wizardry and highlighted the danger of the television medium overriding the choreography (Bayston, 1987; Rubidge, 1993). As Rubidge (1988a) points out, however, the aim of the project was less about creating innovative works than allowing directors, technicians, choreographers and dancers to gain a greater understanding of each other's media. The project was not only an important learning process for all the parties concerned but also provided a vital opportunity to experiment with the relationship between dance and television. The impact of the *Dancelines* programme should not be underestimated, in that it instigated other series that have also dealt with the creative interrelationship between dance and television. The following year a new *Dance on 4* series was commissioned that featured works originally conceived and choreographed for the camera: Darshan Singh Bhuller's *Exit no Exit* (1988), Gaby Agis's *Freefall* (1988) and Yolande Snaith's *Step in Time Girls* (1988).[9] This innovation has continued with the BBC's *Dancehouse* (1992) and *Dance for the Camera* (1994–96 and 1998) series, and Channel Four's *Tights Camera, Action!* (1993–94).[10]

To focus only on film and television as contexts for screen dance would be something of a misrepresentation in light of the current interest in dance and digital technologies. Although this area of artistic research and development is only in its early stages in the United Kingdom, it is worth highlighting some of the key areas of work in this domain. A recognition of the growing concern with dance and digital technologies was reflected in the inclusion of a series of workshop/performances, titled *Techno Dance Bytes*, in the 1995 London Dance Umbrella Festival. This event allowed teams of choreographers, dancers and digital artists to collaborate together on a project of their choosing whether the outcome be a live performance, an installation or a digital format such as a CD-Rom (Keidan, 1998–99; Rubidge, 1999). *Techno Dance Bytes* has been followed up by *Digital Dancing*, an annual platform supported by London Dance Umbrella which provides time, space and equipment for selected practitioners to experiment with ideas and pursue ongoing projects. Yet before highlighting specific examples in more detail it is necessary to address some of the thorny definitions that surround 'digital dance'.

In recent academic literature there is much speculation over the impact of new technologies on our lives (Featherstone and Burrows, 1995). Digital technologies such as CD-Roms, computer programmes, the internet and virtual reality systems are the subject of this discussion rather than analogue technologies such as radio, film and video. The essential difference between analogue and digital systems is that the former is characterized by 'a proportionality between physical changes in a signal and changes in the information it represents', while the latter 'requires a code table...based on physically arbitrary symbols' (Eglash, 1995, p. 18). Yet aside from a technical explanation of 'digital', the question remains as to what constitutes 'digital dance'. Rubidge (1999) argues:

> I would suggest that digital dance must involve the conspicuous use of choreographic concepts as an organising principle, rather than as a means of realising a more general artistic vision. In this way, a 'choreographic sensibility' can dominate a work which bears more resemblance to an installation than a dance work, or a work which does not even feature images or representations of the human or anthropomorphic body...(pp. 41–2)

As already suggested, digital dance can take many forms, but it is that which exists on screen that is of particular interest to this book. To take the computer screen as a starting-point, there are a number of ways in which dance has come to occupy this space. One example is the CD-Rom which, as Jones (1996) identifies, can be used to store vast amounts of information and is often employed as an interactive educational tool. Choreographer William Forsythe has collaborated on a commercially available CD-Rom, *Improvisation Technologies: A Tool for the Analytical Eye* (1999, Zentrum für Kunst und Medientechnologie), through which the user can access information about his choreographic practices. The opening page is like a table of contents and, from this point, the user is able to click on to various titles and images in order to see Forsythe discussing compositional strategies, dancers demonstrating these ideas, and his company performing excerpts of repertoire.

In addition to the concept of CD-Roms as a 'knowledge base' for dance, some artists have begun to explore this format as a purely aesthetic text. Although it is not commercially available, choreographer Mark Baldwin and digital artist Carol Murcia collaborated on a CD-Rom called *Who Killed Me?* (1998). The idea for it is based on a murder scenario and involves six suspects who are colour coded. By clicking on each figure the user can travel into different virtual spaces that reveal disparate images such as rows of bottles, sequences of dance, and small black dogs that race across the screen. Another example is *Windowsninetyeight* (1998) by choreographer Ruth Gibson and digital artist Bruno Martelli, which is described in the on-screen introduction as 'Lo-fi kitchen sink dancing on CD-Rom'. The opening image is a high-rise block of flats and through clicking on different 'buttons' the user can navigate her or his way through graphic images and domestic scenarios. The various pathways through the CD-Rom follow the lives of three women, and many of the images employ an interactive element. By dragging on the cursor the user can activate video sequences, text and sound. For instance, as the cursor moves around the screen various models of washing machine appear, each accompanied by a different soundtrack. Dancing figures run up staircases, roll over armchairs and digital bodies merge and overlap.

A number of computer programmes are currently used by digital dance artists and perhaps the best known for the purposes of dance is *Lifeforms*. This programme was developed by Dr Tom Calvert at Simon Fraser University, British Columbia, and can be utilized for choreographic practices, notation, reconstruction and movement analysis (Jones, 1996). *Lifeforms* consists of a digital image of one or more flexible figures that can be articulated to move in space and time. Clearly, the programme offers a number of benefits in that practitioners can try out choreographic ideas on these inexhaustible virtual bodies thus saving the time and expense of rehearsing with actual dancers (Wyman, 1991).[11] Merce Cunningham and Mark Baldwin are two choreographers who have experimented with *Lifeforms* as a choreographic aid. In addition to *Lifeforms* there are other programmes, such as *Director, Premiere* and *Photoshop*, which can be employed to produce, edit and manipulate images of dance on the computer screen (Anderson, 1996).

As well as specific computer programmes, the internet can act as a site for dance on screen. Since it has become increasingly cheap and easy to create a personal web site, many dance companies and independent choreographers promote their work through the internet. Whereas some of the more basic sites may only have a home page, which offers a small amount of text and perhaps an image of the company or choreographer, those with greater financial and technical support can create complex sites that include video clips of repertoire. Not only can the internet be employed as a promotional vehicle or information system, but it can also act as a performance platform. Dance practitioners can create sites that have a primarily artistic remit and can facilitate an element of interaction with users. The interactive component may include posting responses to the work on a bulletin page or being able to manipulate the digital dance images in some way. As Parry (1998–99) notes, performance on the internet poses questions about the nature of art, authorship and distribution, since images can easily be copied and downloaded, and the global nature of the internet offers access to the user irrespective of geographical locality.

Another digital system that is used by a number of dance practitioners is telematic technologies which can connect images and sounds across remote sites (Hansen, 1998). An obvious example of this is the video conferencing facility that is used in the corporate sector to set up meetings between clients and representatives in distant locations. There are several ways in which this can be achieved. The most sophisticated method is through using ATM (Asynchronous Transfer Mode) which produces rapid and clear connections, followed by the slower, grainier images of ISDN (Integrated Services Digital Network) (Hansen, 1998). There is also CUSeeMe software which can be downloaded from the internet, but this is relatively lowtech and suffers from glitches and time lags (Boddington, 1999). Certain dance practitioners have used these various technologies to differing effects. At the high-tech end of telematics, choreographer Wayne MacGregor linked up dancers in London and Melbourne in *Trial by Video* (1998). The result was an interaction between live performers and projected digital images of the dancers from the remote site. Part of the attraction of this type of work is the choreographic play and potential interaction between bodies that are both present and absent. Similarly, in *Ghosts and Astronauts* (1997), choreographer Susan Kozel used CUSeeMe software to link up dancers in the Riverside Studios and the Place Theatre in London. At the Place Theatre site, Kozel was suspended from the ceiling behind a screen. The audience was free to move around the space in order to observe the live component along with images from both the Place and Riverside. The projected images were of blurred monochrome bodies, interrupted by time delays, which gave a raw and jerky edge to the work.

The technology of motion capture has also intrigued dance practitioners in the digital domain. Parry (1998–99) describes motion capture as an optical or magnetic process of inputting movement into a computer. She states, 'Sensors are attached to pivotal parts of a performer's body: the trajectory of the moving parts is recorded electronically and then transformed into computer-generated animation' (p. 13). This data can then be manipulated on screen and represented through a multitude of visual forms. Digital dance is still very much in its infancy, and as technology becomes increasingly sophisticated there is even more scope in the future for artistic and pedagogic developments in this field. As some of the evolutionary strands of dance on film, television and digital formats are traced in this section, the remainder of the chapter is concerned with the current context for dance on screen from critical, artistic and technical perspectives.

Screen dance and critical perspectives

The translation of dance to screen has been met with both celebration and disdain. Whereas some critics believe dance and screen media to be mutually compatible, others see the two as diametrically opposed. De Marigny (1988) argues, 'The American popular term "The Movies" indicates just how fundamental movement (if not dance) is to the [film and television] forms' (p. 2). In stark contrast, Sacks (1994) states, 'Dance and film are inherently incompatible: film is realistic, dance unrealistic' (p. 24). Critical reviews of dance on screen reveal a particular agenda in terms of the value and belief systems that inform the writing. An analysis of the criteria that critics use to describe and evaluate screen dance can usefully illustrate these hierarchical frameworks and expose the perceptions, concerns and priorities that arise from the relationship between dance and film/television. The critics' attitudes towards screen dance range from a deep distrust and suspicion of the technological mediation of dance through to an acknowledgement that the film/television apparatus plays a vital role in its construction and accordingly champions its creative potential.

The interception of dance by the screen media has clearly provoked strong reactions and some dance critics express an almost technophobic sentiment. For instance, Bayston (1987) states, 'What television dance needs, like every other dance organization, is good dancers and good choreographers, otherwise there is the danger of the medium becoming the message and the choreography smothered with technology' (p. 707). He clearly sides with the features of dance that constitute live performance, such as the dancer and choreographer, and fails to recognize the part that the director and the televisual apparatus play in the creation of screen dance. In a similar vein, Penman (1994) describes the *Tights, Camera, Action!* (1994) series as being 'dominated by ideology and technology. God help dance if Channel 4 has shown us its future' (p. 1173). Implicit within these comments is an adverse reaction to the role of technology in dance. It is perhaps indicative of the critical frameworks employed by these two writers and of the periodicals with which they are associated which underpin Bayston and Penman's attitudes towards the televisual mediation of dance. In the above examples, the two writers were acting as the 'television critics' of *The Dancing Times*, a monthly periodical that focuses almost exclusively on ballet. Hence it could be argued that Bayston and Penman approach screen dance from a classicist hierarchy in which the virtuosic, live dancing body is placed at the pinnacle of any evaluation.

These attitudes towards screen dance are typical of other critics who even question the suitability of dance for the television medium at all. One critic suggests, 'Dance ... is a live thing, and quite how it fits on TV is a moot point' (JN, 1995, p. 70). This statement clearly privileges the 'live body' and sets up a negative perspective for the possibilities of dance on television. Some critics have more specific qualms about the suitability of dance for the screen. In a discussion of dance in 'the movies' Barnes (1985) suggests that screen dance lacks the element of risk that live performance has. Although it is true that any mistakes can always be edited out of dance films, this view neglects to acknowledge the heightened sense of suspense that can be achieved through certain camera angles and editing devices. A similar example can be seen in the way that Parry (1994, p. 12) criticizes one video dance work for its 'distracting' cuts across different locations while praising another for its 'unified setting'. This again reflects a failure to address the television medium and its conventions. Although stage dance is obviously limited to a single setting, the potential to cut from one location to another is a common televisual practice. It is perhaps significant that Parry is a (theatre) dance critic and hence her critical bias is sided with live performance.

A number of critics have taken issue with the content of screen dance, particularly dance that is originally choreographed for the television camera. Within this new and experimental area of work there is a prevalence of everyday movement and pedestrian gesture (see

Chapters 3 and 4) which has provoked adverse reactions from critics. In a review of *Tights, Camera, Action!* (1994), Penman (1994) states:

> There were plenty of people on camera standing about, sitting up, lying down, posturing, gesturing, and posing significantly. There were close-ups on faces, long shots of dockyards, swirling cameras in bull rings, slow mo., cross fades, white-outs, tracking shots, speeded-up time shots, and pixilated images. But dance of any established movement vocabulary was distinctly lacking. (p. 1172)

This type of critique sets up a hierarchical conceptualization of dance, with the virtuosic body and codified techniques at the peak of the evaluation, and only recognizes the explicit movement content as the discernible element of dance. This attitude fails to acknowledge the role of the televisual apparatus in the construction of motion within screen dance. These sentiments, which implicitly suggest that dance can only be defined in terms of trained bodies and established techniques, are echoed elsewhere (Bayston, 1987; Penman, 1995). With reference to the *Dancehouse*(1992) series, Bayston(1992) states: 'The series does raise some points for discussion – where does movement end and dance begin? how far should television tricks support choreography?' (p. 950).

Once again, however, this dismissal of movement techniques that do not conform to existing dance vocabularies highlights the critical framework of these writers. Penman and Bayston's classicist perspective constructs a narrow conceptualization of 'dance' that is located in existing schools of technique and the virtuosity of the performer.

Significantly, some critics, when describing dance that is specifically choreographed for the small screen, neglect to make any reference to the televisual aspects of the work. In a review of *Beethoven in Love* (1994), choreographed by Liz Aggiss and directed by Bob Bentley, Parry (1994) notes:

> [Aggiss] is an operatic dominatrix, her profile as sharp as the shears with which she castrates Beethoven's flowers. The deaf genius is incarnated by Tommy Bayley, eccentric, isolated, vulnerable, trusting that Aggiss, as his Muse, would never betray him. The strength of the piece lies in their quixotic relationship and in the songs composed by Cowie. (p. 12)

There is no suggestion whatsoever of the dramatic framing devices and sharp cuts that also characterize this piece. Similarly, with reference to

Hands (1996), choreographed by Jonathan Burrows and directed by Adam Roberts, Penman (1996) writes:

> A working man sits silently with his hands resting gently on his thighs. In gentle, restrained movements and gestures the man's hands create a lyrical and expressive performance that suggests the skills, intelligence and labour of love required of master craftsmen. There are hints at ballet positions, at sign language and supportive partnerships. (p. 1121)

Again, it is striking that Penman fails to mention the use of black and white and the unedited close-up camera work that are intrinsic to the design and quality of the piece. It is clear from these descriptive analyses that both Parry (1994) and Penman (1996) are conceptualizing screen dance within a theatre dance framework. The focus of their writing is geared towards dance content and emergent themes, without any indication of the role that the televisual apparatus plays in the movement and thematic construction. There is also a notable absence of directorial credits in Penman's review. The fact that he diligently cites every choreographer in the *Dance for the Camera 3* series but fails to mention the directors suggests that, in his view, the choreographic content is a more significant feature of screen dance than the televisual devices. Indeed, this tendency to overlook the televisual components perhaps reveals an underlying snobbery towards television as 'low art'. Meisner (1991) comments, 'There is a lurking suspicion that many of those habitués of the theatre auditorium regard television as a trivial medium' (p. 18).

Some writers have usefully highlighted the difficulties in creating screen dance; yet they do not necessarily present any solutions or alternative approaches. In an article that calls for a symbiotic relationship between dance and television, Schmidt (1991) offers a general summation of dance works created for the screen as:

> examples of pretentious and affected artistry, a festival of running colors [*sic*] and fading images; of abstruse poses and pseudo-elegaic gestures. In place of symbiosis for the mutual benefit of choreographer and film-maker, one finds the dominance of the man (or woman) in control of the camera and blue box, or even occasionally the choreographer being raped by technology. (p. 98)

Although it is perfectly understandable that critics should find fault with particular screen dance works, it is of paramount importance that they should have an evaluative criteria with which to analyse dance that is designed for the camera. It is also fundamental that they take both the dance and televisual/filmic components into account in their evaluative framework, in that screen dance is a fusion of these two disciplines. Through an interdisciplinary approach, critics can then offer some considered guidelines as to what makes 'good' screen dance. As with Schmidt (1991), Penman also outlines some of the problematic issues that surround dance made specifically for the screen; yet he too offers no solutions to these difficulties. In a review for *Dance for the Camera 1* (Penman, 1994) he states:

> the series raises interesting questions. First, was a real partnership achieved between dance and the camera? In the end was the dance dominated by production values – camera work, lighting, editing, sets and so forth – or was dance movement so liberated that the camera followed the dance as it ran and spun and turned? (p. 497)

Penman reaches no conclusions as to what features of dance and television can create effective or successful 'dance for the camera'.

Although there is obviously a fundamental hostility within some critical writings towards the televisual and filmic mediation of dance, several dance critics perceptively argue that if dance is to be created for the screen then it must be altered and shaped to suit the particular medium. For instance, Craine (1995) suggests that dance can make good-quality television if it is especially 'tailored for the electronic medium' (p. 6), and Brown (1994) states that the 'camera and dance have a great deal to offer each other, but dance has to make major adjustments' (p. 10). Likewise, Burnside (1994) clearly supports the creative potential of dance that is conceived for the screen. With regard to the *IMZ Dance Screen* festival (1994) she states:

> Works specifically made for video or stage performances reworked to take account of the requirements of the small screen were the ones which interested me. This meant that it was not wonderful choreography, wonderful dancing or even wonderful camerawork which held the attention. It was some combination of these elements which came together to create something different. (pp. 14–15)

Some critics outline specific ideas with regard to the type of dance that suits the television medium. Brown (1994) suggests that '[television] likes little movements, to crop, enhance and juxtapose' (p. 10), in reference to the potential use of close-up that can be achieved only through the camera. Parry (1994) suggests that 'dance-theatre works better on television than formalist dance' (p. 12), an idea which is also echoed by Meisner (1991, p. 17). Although it can be argued that there have been some equally successful pieces of formalist screen dance,[12] the pertinence of these comments is that both critics are taking note of televisual conventions. The fact that 'narrative' and 'character' are regular features of the television medium, with its potential for close-ups, reaction shots, point of view shots,[13] and multiple locations, is perhaps the reason why the two critics see dance-theatre as being more suited to the medium.

In contrast to the almost technophobic sentiments expressed by some critics, certain writers do acknowledge the extent to which the filmic/televisual apparatus is intrinsic to screen dance, in particular that which is originally conceived for the screen. For instance, Robertson's (1995) description of *Touched* (1995), choreographed by Wendy Houstoun and directed by David Hinton, clearly pays tribute to the way in which the style of filming contributes to the movement quality. He notes, 'Set in a crowded club, lensed in black and white and shot in a series of jump-cut close-ups, it pulsates with an edgy and claustrophobic urban atmosphere' (p. 157). This sensitivity to the role of the camera work and edit is also demonstrated in Mackrell's (1994) evaluation of *Should Accidentally Fall* (1993), choreographed by Yolande Snaith and directed by Ross MacGibbon. She states, 'Suspended in space, [the camera] seemed to hold the lines and curves of the choreography for our contemplation, then diving along the bodies, it caught the movement's vertiginous edge, its heat and grunt and power' (p. 44).

Merrett (1990) even goes so far as to conceptualize the camera in *Scelsi Suites* (1990) (choreographed by Nicole Moussoux and directed by Dirk Gryspeirt) as a 'third member of the cast' (p. 256), an assertion that strongly endorses the belief that the technological apparatus is a vital component of dance choreographed for the screen. Similarly, Rosiny (1994) notes:

> The camera emerges in the latest film works as a sensitive partner in dialogue with the dancers, leads the actors into wider spaces, makes use of its intrinsic mobility – whether on a rostrum, tracked or carried by talented camera operators who themselves move in almost dancerly fashion ... (p. 82)

It is undoubtedly significant that many of the dance critics, such as de Marigny (1990, 1991), Meisner (1991), Mackrell (1993), and Burnside

(1994), who support and praise the recent work in screen dance have also written some key articles that either attempt to outline the fundamental differences and areas of compatibility between dance and television, or have done some research into dance choreographed for the camera through attending various screen dance festivals. They clearly present a considered and well-informed study of the specificities of screen dance.

It appears that screen dance demands a different set of critical criteria from the dance that is performed on stage. Yet there is still a tendency for critics to privilege the codes and conventions of live performance when reviewing dance on film and television. Consequently, such critics conceptualize screen dance from a partial and biased point of view. Such approaches suggest an element of neglect at least and a technophobic sentiment at most. Those critics who do examine screen dance from an interdisciplinary perspective employ a much more balanced framework of analysis. As screen dance is a discipline in its own right, critics need to be equipped with a knowledge of film and televisual traditions as well as of dance. It is only through a dual critical methodology, based on a sound understanding of dance and film/televisual practices, that an evaluative criteria for the study of screen dance can really develop.

A focus on screen dance practitioners

While critics grapple with developing a conceptual and aesthetic framework as a basis for writing about dance on screen, the creative potential of screen dance has attracted a range of eminent practitioners from the fields of dance and film/television. Several well-established directors and choreographers are actively involved in making and promoting[14] dance designed for the camera. Although it is these individuals who are responsible for its creative and practical exploration, there is a dearth of literature devoted to the role of the practitioner.[15] This is paradoxical in view of the number of artists who are keen to explore the symbiotic relationship of dance and screen, especially within the experimental arena of dance created specifically for television. Consequently, this section focuses on the way that a selection of practitioners conceptualize and deal with the televisual mediation of the dancing body. This embraces a diversity of perspectives that range from the creative possibilities that the televisual apparatus can bring to a dancing body, through to the practicalities of performing before a camera. The information for this discussion was obtained through interviews and includes thoughts and ideas from seven practitioners who have worked on the *Tights, Camera, Action!* and *Dance for the Camera* series. These individuals were selected because of their experience with, and commitment to, work that sets out to explore the creative relationship between dance and television. The interviews took place between July and October 1996 and involved directors Margaret Williams,[16] Tom Cairns[17] and David Hinton,[18] choreographer-director Alison Murray,[19] choreographer Wendy Houstoun,[20] and dancers Emma Gladstone[21] and Anna Pons Carrera.[22]

In contrast to some of the negative attitudes expressed by dance critics towards screen dance, the above practitioners are keen to commend the potential compatibility of dance and television. Hinton notes, 'Dance is obviously a wonderful thing to film because it's all action and motion, which is what film should be', and Williams says of her initial experience of directing dance, 'This opened up for me a whole new way of looking at things and a whole new interest in terms of what I could do within the film frame'. It appears that one of the immediate considerations in making dance for screen is the creative possibilities that the television medium can bring to dance. All of the directors and choreographers suggest that, to some extent, televisual considerations are a vital starting point for each piece of work. Hinton states, 'What works well on film or on television is completely different from what works well in the theatre. So is it possible to create something that is from the outset purely cinematic and not theatrical?'

Similarly, Williams says of *Le Spectre de la Rose* (1994) (choreographed by Lea Anderson), 'I liked the idea of manipulating the film in some way, so that the movement would be changed'. Cairns was equally keen to employ the television medium to the full: 'the brief had always been for the [*Dance for the Camera*] scheme to try and make films specifically for the screen, and not just to film dance, i.e. glorified live performances on film.'

It therefore seems that part of the agenda of dance designed for the camera is to avoid the transposition of a stage context to screen. One way in which this can be achieved is to locate the performance in a site that would not bring to mind, or normally be associated with, a stage setting. For instance, Williams chose to film *Le Spectre de la Rose* in and around the grounds of Nymans Gardens. She explains, '[I] have always been keen not to film something that has a flat back wall, i.e. what you get on a stage'. Similarly, Murray's Kissy *Suzuki Suck* (1992) is set in a car, and *Alistair Fish* (1994) and *The Storm* (1992), by Cairns and Aletta Collins, take place aboard a train and in a restaurant, respectively. Yet, it appears that to relocate dancing bodies into a geographical site that cannot be achieved on stage does not always disassociate the work from a stage context. Hinton argues that it is easy to fall back on theatrical devices: 'Lots of dance films take the action out into the world. I mean they do it round a swimming pool or in a field, but actually the language, the movement itself tends to remain very theatrical.'

Implicit within the practitioners' comments is the need to comprehend the fundamental differences between what characterizes the television medium and what typifies stage performance. What suits one medium does not necessarily suit the other, and as Hinton suggests, 'the dance that is most effective in the theatre is not necessarily the dance that is going to make the most effective films'. This has certain implications in terms of how the dancing body is represented and employed in screen dance. For instance, when creating *Touched* (1994), Houstoun noted, '[the movement] was only going to work if it worked within the frame, not if you looked at things with your naked eye'. One of the essential differences between stage and screen is with regards to 'time'. This is significant in terms of how long a choreographic idea can be explored and the speed at which a movement is executed. Houstoun became very aware of this issue with *Touched*. She states, 'There's a different sense of time [in television], and that's how little you need to establish something, and how quickly information reads'.

There is also a need to consider the different characteristics of a 'live body' as compared to a 'screen body'. For instance, at the point at which it becomes a screen body, the live body goes through an immediate metamorphosis. Murray suggests:

> I think people make the mistake of thinking that if they film something with an uninvolved camera, with natural lighting, that it's going to seem very real. But as soon as you frame something within the screen it's no longer real anyway and viewers bring a different set of expectations to the screen.

To a certain extent this is tied up with the simple act of framing a body and then relocating it into the context of a two-dimensional screen: the body inevitably becomes distorted. Houstoun remembers how the live body takes on a completely different quality in a screen context: '[the image] just looked much more dense. I remember, when I first saw the monitor, getting really excited ... you don't imagine what lighting and the quality of film will do to the image.'

It is notable that the directors all referred to visual forms, in particular film and television genres, with regard to influences on their work, rather than to theatre dance traditions. For instance, the strong sense of design and composition that characterizes Williams's work owes to her visual arts background. She notes: 'I trained as a painter...my films come out of storyboards and sketch books, and images that I've collected – I've got rafts of notebooks that have pictures in that have influenced me. Design is a very important element of the films.'

Likewise, it is evident that, for Murray, screen images have a major influence on her work. She suggests: '[My influences are] definitely more from the pop cultural end of the spectrum. Some people say pop promos and MTV [are] not art, but I think that a lot of things that are being done in commercial territory are actually more ground breaking than a lot of high art developments...what else? comic books, animation!'

Hinton is similarly influenced by specific film genres, particularly in relation to their treatment of moving bodies. He states:

> I like films like *Raging Bull*...the boxing sequences in it are fantastically powerful movement sequences that are intensely cinematic. The power of the way in which they're presented is not just to do with the action that's going on, it's the way it's shot and edited. Another thing I really love is Kung Fu movies. They're purely cinematic in the sense that they use the resources of what film can do to present what are in fact physically impossible actions.

Hinton's fascination with the possibilities of movement on screen does not derive from a dance tradition, but from popular action films. This perhaps suggests a blurring of boundaries through the way that other visual influences inform the practitioners' work.

It was suggested earlier that many of the practitioners are keen to construct movement images that employ the television medium and to avoid recreating a stage context on screen. One approach is to employ standard televisual conventions in the creation of screen dance. With *Alistair Fish* and *The Storm*, Cairns was keen to construct dance that could be inscribed with common televisual devices. He notes:

> I wanted to apply the same techniques to the way I filmed [them] as I would do for a drama or anything else. So I was quite concerned to do close-ups, and two-shots, and wide shots and cut it together like I'd cut anything...That was a very deliberate decision. I didn't have a specific dance mentality when I was making [them].

One of the most striking components of screen dance is the relation-ship between the body and the camera. There are clearly many vari-ables in terms of where and how a director may position a camera in relation to the body and it is evident that the camera is an essential element of the choreography. Hinton suggests:

> Any movement of the camera itself is more powerful cinematically than the movement of a person in a fixed frame. So the movement of the camera, in what I would call a cinematic kind of choreogra-phy, has to be an element that you take into account.

Similarly, in reference to choreographing *Horseplay* (1995), Murray states:

> I just wanted to explore the dynamics of female playfulness and I wanted the camera really to be a part of that. So I ended up using a minicam strapped to a boom pole so that it could go in and out and all around the dancers. I really choreographed a lot of the material based around the location and that particular camera.

The conventions of camera work can either be employed or aban-doned. For instance, Murray suggests, 'If you want someone to be mov-ing across the screen, and you want to see that movement across the screen, don't pan the camera because [the person is] going to stay in the centre of the screen.' Yet although the practitioners are aware of how televisual conventions can effectively complement the moving body, they are equally happy to exploit and manipulate existing tradi-tions. Cairns states:

> I think, like anything, you can and sometimes should break the rules or set-up...we cut an awful lot on 'the move' as they say. You know, we cut on a movement into another movement and that's very different to cutting from static shot to static shot, or someone just leaving a frame and then someone walking into a frame.

Similarly, Murray refers to 'cutting down the line':

> When people talk about cutting down the line, that means that if you've got a wide shot, you should change your angle if you're going to go into a tighter shot, whereas I'll just zoom the camera in and then cut, and then in again. If you know the rules then you can break them.

Editing is a feature of the film and television media that cannot be reproduced on stage and, as with camera movement, is an essential component of screen dance with regards to rhythm and motion. The edit delineates the televisual structure and the way in which a dance piece is edited can clearly affect the pace of the work, its sense of shape, and how the images are read. Williams is extremely concerned with the quality of the edit and the extent to which it can manipulate a work. She states:

> For me, the edit is what makes the film. I find cutting the most exciting thing imaginable, because you're in a situation where, actually, you can do whatever you like with the material that you've got. Two people given the same material (rushes) to edit would end up with completely different films.

In terms of editing, there is also a different sense of time between the stage and television context. Murray comments:

> The way a rhythm might work when you're dancing live or on stage doesn't mean it's going to translate to the screen and one thing I find is that I cut sooner than I used to. I realized that you don't need to see as much and that your mind fills in the gap.

The differences in time between the dance and television media were something that Houstoun was unprepared for:

> Often, what I'd done was imagine one shot would lead into another – in my mind I would have edit points...but actually, they didn't work. Or the sequence was too long so, although the edit point was right, it should have actually been crushed into three seconds.

In addition to the technical considerations that inform the televisual mediation of dance, there are also several implications for the 'performing body'. One of the notable factors that arises from video dance performance is that dancers are being required to execute styles of movement that would not be appropriate for a stage context. For instance, Gladstone comments:

> That was weird movement [in *Sardinas*][23] because we were on our sides pretending to walk and somersault. That was a completely different kind of movement than we'd ever do on stage. You could only ever do that on camera which I think is why it works so well.

Similarly, Carrera states, 'When we are doing [facial] close-ups, the choreography is actually the [facial] expression, which is never like that in the theatre'. As well as different movement demands, dancers are also required to modify their performance style in order to suit the subtlety of the camera's scrutinizing gaze. Houstoun observes, '[If a dancer] is so expressive, it's too much. The camera can't take it, especially in close up'. Yet this is something that performers can explore and experiment with. In relation to *Touched*, Gladstone remembers, 'I suppose as we went on, I became more aware that you can play with [expression] more'.

During the production process, there are many practicalities that, in one way or another, create difficulties for the 'performing body'. One of the features of screen work is that the dancer may have little comprehension of how the completed work will appear. Carrera comments: 'When you're rehearsing for the theatre, you see the continuity of things. When you are rehearsing for a film, if you are not the choreographer or the director, you lose a bit of the [sense] of what's going on.'

There are also tremendously long gaps throughout the filming day, which can be equally problematic for the body. Carrera observes, 'You feel like you've done nothing and you've been hanging around a lot'. There are so many technical considerations when filming dance that, at times, it seems as if the actual body is the least important factor. Gladstone notes:

> With Wendy [Houstoun's] work, props are often just deadly... or you're dancing on gravel outside and it's freezing cold. I almost feel like the movement's the least important thing... the medium takes a lot of effort from so many other people to make it happen.

The television medium appears to put a whole set of different demands and strains on the dancing body. Gladstone feels that when it comes to film work, she loses the sense of control that she has in stage performance. She states:

> The practicalities of the intrusion of the camera and the film crew and everything else I found very constricting, and it took quite a long time to realize that actually I had to not be as precious as I would choose to be on stage about my dancing, my performing in it.

It is clear that the dancer's body has a completely different set of performing experiences for screen dance, from the one it would have for a stage performance.

It is suggested earlier that several critics conceptualize screen dance within theatre dance criteria. These attitudes, which completely bypass the role of televisual apparatus in the construction of screen dance, filter through to practitioners who are clearly frustrated by such misconceptions. Williams comments:

> There is a feeling in the dance press that the choreography is the most important thing and the choreographer makes all the decisions. It's time for the dance press to look at these films from a new and informed perspective.

Murray is equally concerned by the prioritization of theatre dance values that implicitly dismiss alternative movement styles. She argues:

> Dance on the screen, to me, is a completely different genre, field, art than dance on the stage, and the technology is part of it. Dance critics [who view it within a stage criteria] are completely ignoring the whole other medium of film and television, which has its own aesthetics and own conventions completely separate from that of dance on the stage.

Yet those practitioners involved in the genre remain committed and inspired. Williams enthuses, 'You can [film movement] in so many different ways... it's tantalizing.' Likewise, Hinton proposes, 'In a way, it's incredibly exciting... [dance for the camera] is an idea the potential of which has hardly begun to be explored.'

The practitioners draw attention to a number of significant features and components of screen dance: the specificities of working within the television medium; the creative possibilities that dance can bring to television; the role of the camera and the edit; the influence of location; and the creative tension that can emerge through interdisciplinary work. These perceptions provide a valuable insight into the concerns and ideas that are tied up in the creative production of screen dance. Many of these ideas are returned to in Chapters 3 and 4, which focus on dance conceived for the camera. The final section of this chapter, however, looks to the technical framework of dance on screen.

The live body and the screen body: a technical comparison

The presentation of a 'live body' is unavoidably transformed when it becomes a 'screen body'. The distance between the camera and its subject matter, the angle and focus, the use of colour and lighting, and the style of editing all contribute to this modification. This undoubtedly has far-reaching implications for dance, in which the moving body, with its particular spatial, temporal and dynamic characteristics, is central to the aesthetic. Therefore the aim of this section is to make a comparative analysis of the live body and the screen body, the theatre setting and the television context. Although Chapter 1 makes reference to both film dance and digital dance, this particular section focuses on the medium of television in order to provide the groundwork for the case studies that follow. In Chapter 2 the majority of examples are of dance that is seen on television. The one exception is the Hollywood genre of late 1970s and 1980s dance films, which were originally made for the cinema but are now regularly screened on television. Meanwhile the subject of Chapters 3 and 4 is dance that is originally conceived and choreographed for the camera and, again, all of these works were designed for television. Thus in order to provide an analytical framework for these examples, this section compares the fundamental differences between the live body and the screen body within the contexts of stage and television.[24]

Dance is characterized by its use of space, time and energy in relation to movement and it is these three phenomena that can be most distorted through the television medium. To begin with the concept of space, one of the key characteristics that differentiates the stage setting from the television context derives from the way that the human eye perceives space and the way that the camera perceives space (Lockyer, 1983). When a spectator looks onto a stage, the space generally takes the shape of a rectangle.[25] The law of perspective creates an impression that the stage is slightly wider at the front and then tapers inwards to become marginally narrower at the back. A camera perspective, however, shows the reverse. The field of vision of a camera is coneshaped; the space is extremely narrow at the front and then extends out to become widest at the back. Consequently, the perspective of a camera sets several limitations and potential possibilities for the choreographic design. For instance, a dancer who is close to the camera can take one step to move out of view, while a dancer several metres directly behind has to take many steps to move out of the frame. Similarly, during stage performances the dancers can usually only enter and exit from side wings, but within the camera space it is feasible for dancers to enter and exit from behind the camera so that they appear to enter from 'camera front'.

A number of differences also exist between the television 'screen space' and the space that the viewer perceives on stage. One of the foremost distinctions is the literal shift from a three-dimensional stage to a two-dimensional screen. Although, to some extent, the Western eye has been trained to perceive three-dimensionality within two-dimensional images (Monaco, 1981), televised dance has a certain 'flatness' that is uncharacteristic of stage dance (Newman, 1985; Maletic, 1987–88). Another spatial feature is the difference between the size of the television screen and the theatre context. Whereas a live body is in the region of five to six feet tall, a screen body in 'full shot'[26] is roughly 12 to 16 inches tall. The result is that the detail of a live body is automatically diminished when it becomes a screen body in full shot. Yet regardless of this specific limitation, through the use of close-up, the camera is able to focus on detail that the naked eye would not be able to perceive. For example, a twitching jaw muscle or the tip of a finger can be enlarged to take up the whole of the television screen. In addition to close-up perspectives there are several other spatial relationships between the camera and the dancing body that could not be achieved on stage.

During live performances the viewer conventionally watches the dance from a fixed upright position, which is face-on to the movement. Aided by the camera, the spectator is able to see the dance from a multitude of angles and distances. For instance, the dancing body can be framed in a big close-up, or from an extreme long shot. It can be filmed from above, from below, and from any particular side. Nor is the dance limited to one particular location. Unlike stage space, televised dance can cut from one geographical location to another. There is also the possibility of camera movement even if a dancing body is static, an impression of travel can be constructed as the camera moves around the body.

In much the same way that the televisual apparatus is able to construct spatial relationships that could not be recreated on stage between a dancing body and a spectator, it is also able to manipulate temporal factors. The screen body can move at certain tempi that the live body could not replicate and this technique can be achieved either during filming or in post-production. For instance, fast motion can be employed so that the screen body moves at an impossibly rapid pace, or the opposite is slow motion so that temporality is elongated. Although of course it is possible for dancers to imitate slow motion, simply by moving at a very slow pace, they nevertheless remain located within the confines of 'real time'. Many movements are not possible at an extremely slow pace. For example a live body can not leap through the air in slow motion in the way that a screen body can.

A similar concept is with the 'still frame', in which an image is frozen, or 'reverse motion', in which the image runs backwards. With many dance movements a prolonged moment of stillness or the reversal of a particular phrase is simply not possible.

Aside from the speed and direction at which a recorded image is played, the other temporal factor that characterizes the televisual apparatus is in relation to editing. One of the most established conventions of film and television is that a particular 'event' is rarely filmed as one continuous shot, but is made up of many shots, which are then 'spliced together'. This process is known as montage or editing.[27] The implication of this device is that time is no longer restricted to linear progression and, as a result, television has the capacity to restructure a passage of events completely. The convention of editing has largely been necessitated through narrative films, in which the occurrences of the plot span several days, months or years. It is clearly not possible to recall these events in 'real time'; hence, the camera simply cuts out the time that is not intrinsic to the narrative. This convention has evolved and become established to such an extent that to see a piece of unedited filming can appear extremely alien to the eye.

The implications of the editing convention in terms of a dancing body are far-reaching in that temporal organization is central to the dance aesthetic. To some extent, editing practices and possibilities can be seen as both a restriction and liberation for the dancing body. One of the complexities of editing a dancing body is that the choreographic design may become distorted as the image cuts from shot to shot. This is clearly a concern when stage performances are adapted for television. One of the most prominent conventions of editing, within both film and television genres, is a practice known as 'invisible editing', which aims to cut from shot to shot as unobtrusively as possible (Metz, 1975; Monaco, 1981). Monaco (1981) describes the way in which 'dead time' may be erased from a scene through the device of the 'cut':

> The laws of Hollywood grammar insist that the excess dead time be smoothed over either by cutting away to another element of the scene ... or by changing camera angle sufficiently so that the second shot is clearly from a different camera placement. Simply snipping out the unwanted footage from a single shot from a single angle is not permitted. The effect, according to Hollywood rules, would be disconcerting. (p. 184)

Although Monaco is particularly concerned with Hollywood practices, the convention of 'invisible editing' has slipped into numerous forms of film and television. Consequently, when filming a dancing body, a single shot of the whole event would appear alien and unconventional to the television viewer; yet, to cut to another piece of action, or to a different perspective of the same piece of action, runs the risk of distorting the overall choreographic shape. This dilemma has caused much debate among the dance and television world (Nears, 1987; Mackrell, 1993) and some of the ways in which these issues have been tackled are examined in Chapter 2.

Although, to some extent, editing practices are perceived as setting up limitations on the dancing body, they also provide several possibilities which enable the screen body to achieve feats that are simply not feasible for the live body. Indeed, several dance practitioners have attempted to employ screen devices in their stage performance. For instance, Pina Bausch often constructs her work through a 'montage' structure in which short episodes, which have no apparent connection, are placed back to back. Yet ultimately, she is working within the confines of 'real time'. Vignettes of apparently unconnected material may be placed side by side but, unlike television, the physical transition from one section to the other cannot be erased in stage performance. This would suggest that in television there is much potential to explore the temporal construction of the screen body: dance phrases may be restructured in several different orders; bodies are able to switch from one geographical location to another; movement phrases may be precisely repeated many times over; and bodies can appear from nowhere.

The extent to which a camera can alter the spatial and temporal configurations of a dancing body is relatively concrete because, technologically and perceptually, it is possible to explain these phenomena. It seems, however, that the way in which a camera affects 'energy' is less quantifiable. Several writers have noted that the camera tends to 'dull' or 'flatten' the sense of dynamic quality that a movement employs, but there appears to be no single explanation for why this transformation occurs (Maletic, 1987–88; Mackrell, 1993). One of the central factors is simply the absence of full-size, live bodies on television. As the television screen provides a diminished and relatively poor-quality image, the detail of a live body is reduced. Consequently, the viewer is able to see less of the effort that underlies a movement and the subtle contraction and relaxation of the specific muscles that this requires. Although this type of muscular detail may be seen on screen in close-up, the way in which the use of energy is dispersed across the complete body is then lost. Due to the 'immediacy' of bodies in live performance, audience members are sometimes said to respond through what is described as 'kinetic empathy' (Hanna, 1988a). This is caused by a particular physical sensation, often associated with the nervous system, in which the body empathizes with the distribution of energy produced by other bodies in motion. The way in which the television spectator is distanced from the presence of live bodies, by way of the television screen, is said to prevent this sense of 'kinetic empathy' from taking place (Mackrell, 1997).

It is also notable that television is highly fabricated. In stage performances, in which the dancing usually takes place in the space of a few hours, the viewer is used to seeing bodies that sweat and breathe heavily and rapidly, especially during sequences that demand a high energy output; whereas in television, dance is filmed in short 'takes' and in between times the performers are 'touched up' by make-up artists and able to catch their breath. As a result, the television viewer sees a distorted sense of effort. Of course, an illusion of energy may be produced by the television medium in other ways, and this is addressed in Chapter 2.

The construction of a screen body is not only dependent on the televisual apparatus, but on the given choices of a particular individual. The transformation of a live body into a screen body is always mediated by way of a person, usually in the form of a director. However minimal the individual's influence, the apparatus of the television medium requires that certain directorial choices must be made. For instance, even if the director simply chooses to switch the camera on and off from a static position, she or he has already brought considerations of distance, angle, shot size, focus, lens type, lighting and shot length into play. In fact, due to the many variables involved in transforming live bodies into screen bodies, a whole selection of individuals may be involved to a greater or lesser degree. In the case of screen dance, this may include the choreographer, a producer, lighting designers, camera and sound operators, a design team and so on. The choices that a given individual has, however, are largely dependent on financial and technological factors. For instance, the number of possible shots may be limited by an overall budget and the type of shots may be restricted by the location. Individual perception is also paramount to the way in which a dancing body is 're-presented'. The various features that one director perceives as being central to dance may be radically different from another director's 'vision' or 'interpretation'.

The decisions of an individual in relation to the parameters of the televisual apparatus are the fundamental factors that determine the way in which a dancing body is re-presented on screen. To some extent the television medium allows for a more versatile construction of a dancing body, since the screen body is able to perform feats that would be impossible outside the spatio-temporality of film and television, but in other instances the dancing body is restricted by the televisual apparatus due to the 'flattening' of space and energy. Many of these issues are addressed in Chapter 2.

The aim of Chapter 1 is to provide a contextual framework for dance on screen. From the above sections, it is apparent that screen dance consists of many forms, genres and styles The dance that we see on screen clearly belongs to a specific historical, cultural and economic context. From the late twentieth century through to the turn of the millennium, there has been a prevalence of two-dimensional representations that have constructed a culture of spectacle or image. Dance is very much a part of this visual framework. Yet dance on screen also belongs to a political and economic context. The work that is screened is subject to broadcasting policy, modes of censorship and financial backing. The evolution of dance on screen is constituted through several historical strands. Developments in film, television and digital formats are compartmentalized though distinct technologies but intersect in their exploration of dance images. From a critical perspective, screen dance has called attention to the framework with which writers analyse and evaluate work. There are some critics who use stage criteria to evaluate screen dance and overlook the role of the film or televisual apparatus. This is clearly a problematic mode of analysis in its partial point of view. Those writers who champion screen dance have had to reconsider their critical framework in order to take account of the technological components. Practitioners involved in making dance for the screen identify the difficulties of working in an interdisciplinary way, but enthuse over the creative potential that derives from experimenting with the dance/screen interface. It is apparent that the screen body has a completely different set of characteristics to the live body on stage. Consequently, inherent tensions surface in the adaptation or creation of dance for the screen. This contextual framework now sets the scene for Chapter 2, which addresses some specific case studies of dance on screen.

2
Images of Dance in the Screen Media

It is suggested in Chapter 1 that there is a rich diversity of dance on screen. These images, in all their different forms, contribute to shaping how dance is perceived both by scholars and students of dance and by the general public at large. The dancing bodies that we see on screen are constructed through the film and television apparatus and different technical and aesthetic approaches are employed in order to create particular representations of dance. Yet these are not simply bodies inscribed with technical devices: they are also bodies that carry social, cultural, political and economic meanings. Therefore the way in which dance is filmed and the moving body as a locus of meaning are pertinent factors in relation to how we read dance on screen. In response, this chapter draws on a selection of case studies in order to investigate critically a variety of ways in which the screen media have dealt with the dancing body. As mentioned in the previous chapter, these are all examples of dance that have been broadcast on television, although one set of works was originally made for film.

The first three case studies are of work located at the commercial or popular end of screen dance: 1980s Hollywood 'dance films', television advertisements and pop music videos. The remaining two case studies are rooted in an art dance context: stage dance translated to the screen and early examples of dance created for the television camera. I have purposely employed a number of different approaches with which to examine these works. In each case, certain factors are always addressed, such as the actual movement content and the ways in which the televisual apparatus participates in the construction of the dance. But I also pick up on other themes that are pertinent to specific examples and which re-emerge later in the book.

Hollywood dance films: popular representations of dance

Representations of dance within commercially oriented films can provide an illuminating case study through which to address the dancing
body as a 'popular image'. At the end of the 1970s and throughout the
1980s, a whole series of Hollywood dance films developed that
reflected the general exercise craze of that era. Films such as *Saturday
Night Fever* (1977), *Dirty Dancing* (1987), *Fame* (1980), *Flashdance* (1983),
Staying Alive (1983) and *Footloose* (1984), which coincided with the
marketing of fitness videos by star personalities and the appropriation
of dance wear and 'workout clothes' into mainstream fashion, clearly
mark the explosion of dance, aerobics and fitness classes that prolifer-
ated throughout the decade (McRobbie, 1990; Buckland, 1993). The
popularity of these films is reflected in financially successful box office
figures (Gow, 1983) and their appeal continues as they have since been
relocated into the domestic viewing context due to home video availability and regular screenings on television.

The 1980s is a particularly interesting period in terms of society's ideas
about the body and very much reflects a notion of the 'body as project'
(Shilling, 1993). This conceptualization propounds that the body is not
an entity which is fixed or stable but is a fluid phenomenon that can be
managed, maintained and altered. Shilling (1993) describes it as a symbol of self-identity, and accordingly the individual has the power to
deconstruct and reconstruct its numerous discourses. For instance, the
physicality of the body can be shaped and modified through exercise
regimes, such as aerobics and weightlifting, or through cosmetic surgery
in the form of breast implants and liposuction. In many ways these
ideas mirror the political climate of the 1980s in 'Thatcherite Britain'
and 'Reaganite America', which promoted the individual's 'right' to
achievement and the potential of the 'entrepreneur' to create and maintain personal success. It is for this reason, that I want to focus on the
1983 film *Flashdance* in particular. The narrative of *Flashdance* follows
the fortunes of Alex, a young woman who aspires to win a place at the
prestigious local ballet school. The following analysis focuses on two
specific dance sequences in the film: the first is of Alex dancing at a club
and the second depicts her rehearsing in a studio. The aim of this study
is to expose how the filmic apparatus constructs a representation of the
dancing body that can appeal to a mass audience and to identify some
of the discourses that are inscribed within this popular image of dance.[1]

The lead-up to the first dance sequence takes place during the opening credits in which images of a steel mill are depicted and the jocular

banter of male voices can be heard. The environment is dark and grimy and, in between images of men hammering, banging, drilling and planning, the camera repeatedly focuses on a welder diligently at work. These representations, in which the welder is firmly placed, construct a notion of the 'labouring body' as tough, masculine and active. It then comes as a surprise when the welder removes 'his' helmet to reveal that this is Alex, a beautiful young woman. The action switches to Mawby's club where Nick, the owner of the steel mill, has been invited for a drink by one of his employees. As the two men are served drinks at the table, some music strikes up and the action cuts to a long shot of an empty stage area.

The stage is black, with a walkway that extends out to the audience, and is framed in red neon light. This classic cabaret setting places the dance within familiar theatrical imagery. After a moment or two, a woman dressed in a trouser suit and stiletto heels struts sexily onto the stage in silhouette. She kicks her leg sharply into the air to execute a half-turn, continues to traverse the stage in a 'robotics' style walk, and then kick-jumps her legs behind her to fall directly to her knees. The movement suggests skill, virtuosity and risk. The shot cuts away to a close-up of the two men, who sit entranced by her performance, and then cuts back to a long shot of the dancer, which implies that this image is from the men's 'point of view'. Her routine continues, drawing on elements of jazz, cabaret dance and popular 'street styles' of the 1980s, which perhaps suggests that she has no formal dance training. Although the dancer is in fact Alex, as yet the spectator is given no indication that it is actually her.

The camera briefly cuts to a close-up of the dancer's feet to highlight her intricate footwork, but then returns to a long shot to reveal that she is now in a scanty red dress. She walks provocatively towards the audience and moves to a chair, on which she writhes, undulates and arches her back. The imagery of the routine has now shifted from the cabaret vernacular into the realms of erotic dancing or striptease; this is apparent both from the obvious removal of garments and the more explicit sexual content of her movement phrases. At this point a woman's voice sings over the backing track, 'He doesn't mean a thing to me, just another pretty face to see', and these lyrics serve a number of purposes. First, they suggest that Alex is indifferent to the men in the bar, which is reiterated later when she pours a drink down the trousers of a man who tries to squeeze her buttock. Second, it refers to Nick, the mill owner, who is sitting in the audience; later he becomes her lover but at present he is simply 'another face'. But finally it implies that this indifference to men is proof of her total dedication to dance. To some extent this validates the fact that she is performing in a sleazy working men's club; the implication is that were it not for her passion for dance (which is tied in with economic necessity as she is saving up to go to ballet school), she would not work here. The shot meanwhile cuts to a close-up of her feet, which maintains the play on sexual imagery. The fragmentation of the erotic body, as seen in the phallic stiletto heels, is a common pornographic convention (Kuhn, 1985). This theme continues as the shot then switches to a close-up of the dancer's face with her thick, red lips slightly parted as she gazes out to the audience: this 'come-on' construction is also a typical image within soft pornography (Kuhn, 1985).

This is the first time that her face has been shot in close-up and it is here that the spectator is given a second surprise: the dancer is revealed to be Alex, the welder. Again, the shot briefly cuts across to the two men, still entranced by her performance, and then returns to a full shot of her body. As the music begins to 'rev up' her movement grows wilder, to include multiple spins, sharp kicks and pelvic isolations. On one level this could symbolize her 'passion' for dance, but the sexual imagery perhaps also signifies an erotic passion, reflected both in the men's desire for her and her love affair that occurs later in the narrative. She goes on to remove her dress under which is a tight red basque. The shot then cuts to a close-up of her stiletto heels digging into the floor and on to a close-up of her hand as it reaches to a chain above her head. Just as she 'pulls on it' (like a symbolic act of masturbation), the image switches to a long shot, showing her in profile, arched back onto the chair, as she is drenched by a great shower of water. This crude allusion to the male orgasm is further heightened when the male audience cheers at the moment of 'ejaculation'.

The sequence continues with her wildly pounding the chair, gyrating on the spot and strutting down the walkway towards the audience. The camera now follows her from behind, and the image loses its ordered red and black clarity of the stage to become a hazy blue wash of smoke-filled lights and sprawling audience. This chaotic image ties in with Alex's post-coital look: she appears drenched and dishevelled as she wildly tosses her head from side to side. At intermittent moments, the shot cuts to the two men in dialogue in which the employee reveals to Nick the true identity of Alex. The final shot of Alex gyrating on the spot cuts to a close-up of a hamburger sizzling on a grill-plate, which creates a memorable montage: comparisons may easily be made between the 'fired-up', naked flesh of Alex and the 'dead meat' as it fries.

In many ways, this short dance sequence offers the spectator a wealth of narrative information. To commence with, Alex's working-class origins are clearly apparent: her working in the seedy men's club; her reliance on popular dance styles rather than high art dance; and the prevalence of the colour red and beads of sweat that fly off her allude to the 'sparks and flashes' of her welding job. The fact that Alex must rely on this 'sleazy dancing' implies that 'art dance' does not traditionally belong to the working classes. There are also suggestions of her future affair with Nick, her indifference and implied independence of men (although this is not strictly the case) and her somatic passion, both in terms of her erotic self (several sex scenes take place in the film) and her complete commitment to dance. The routine meanwhile is a metaphor, in form and content, for the sexual act: the removal of garments, the use of close-ups as a reference to pornographic conventions, the images of masturbation and penetration, the moment of climax and symbolic ejaculation, and the allusions to the post-coital scenario.

To some extent, this filmic construction reflects the puritanical discourses that promote the dancing body as sexually unleashed. This stereotypical conceptualization has been used to devalue dance through the body's links to tactility, sensuality and sexuality, which early Christianity deemed sinful and immoral (Burt, 1995; Hanna, 1988b). Although this sequence does not explicitly denounce the association between the body and sexuality, the way in which it clearly draws on this convention perpetuates the popular image of the dancing body as a site of provocative sexuality.[2] It could also be suggested that the film nevertheless implicitly devalues the dancing body in that, although it is about dance, the sequences are laboured by an abundance of close-ups, numerous reaction shots and repetitive choreography. As Billson (1995) states,' this flashdancing stuff appears strikingly unoriginal (even if she can do 38 pirouettes on her bum); and tricksy filming prevents the routines from ever properly letting rip' (p. 248).

A feminist reading of *Flashdance* might suggest that through the use of narrative conventions and cinematic codes, Alex is constructed as a site of erotic spectacle and is subject, at least within the diegesis,[3] to a male gaze. An alternative feminist analysis could meanwhile argue that Alex is an independent heroine who uses her sexual and labouring body as a form of empowerment: through erotic dancing and welding, she is able to exploit men sexually and financially in order to pursue her dream of becoming a classical dancer.

Within this sequence it would seem that any attempts to be innovative, in terms of either choreographic practices or filming techniques, are marginalized in favour of narrative progression and erotic imagery. Instead the filmic apparatus and dance content are used to construct stereotypical images of dance, as seen through notions of virtuosity and determination, and to set up a sexual relationship between Alex and the male protagonist/spectator. Although the links between the dancing body, class and sexuality continue throughout the film (McRobbie, 1990), it is worth looking briefly at the 'Maniac' dance sequence in which other conceptualizations of the dancing body are constituted.

The 'Maniac' routine is the second major dance sequence of the film and takes place just after Alex has arrived home, exhausted after a tough day's work at the steel mill. The mood and dynamic changes suddenly as the shot cuts to a close-up of Alex's foot as she bandages it with sticking plaster and the upbeat chords of the song 'Maniac' strike into action. The focus on the care and preparation that a dancer's body demands continues as she wiggles her toes in time to the music. This perhaps reflects the notion of 'body as machine' in which the body is conceptualized as something 'to be finely tuned, cared for, reconstructed and carefully presented' (Shilling, 1993, p. 35). The shot switches to Alex's legs as she jumps down from the seat and revolves a couple of turns to test out the bandaging. The majority of the sequence consists of a series of close-ups that repeatedly cut to and from three particular body areas: the buttocks and thighs, the head and the feet.

The close-ups of her buttocks and thighs show Alex in profile as she jogs or undulates on the spot. As she is wearing a high-cut leotard, the majority of the screen is taken up with nude[4] flesh. Once again this places the dancing body within the realms of sexuality and, although in this instance there is no apparent male audience, the close-up camera work is similar to the 'male gaze' of the previous dance sequence. The prevalence of flesh is also typical of pornography films when images of thrusting buttocks take up the complete frame. As the camera cuts to a close-up of the feet, Alex jogs rapidly on the spot. In this instance, dance is constituted as high-impact, repetitive exercise and bears more similarities to the aerobic workouts of the 1980s than to the type of expressive qualities that can be conveyed through dance. This again reflects notions of the 'body as project' (Shilling, 1993) in which the individual works towards maintaining a certain physique. The close-ups of Alex's face meanwhile capture her sweating and panting as she executes this rigorous drill. With her damp hair and driven look, she appears closer to a sportswoman than a dancer. Her face conveys none of the expressive qualities that may emanate from dance, but instead refers to the 'burning will' or 'true grit' of a sports workout. In many ways this ties in with the 'no pain, no gain' philosophy that exists in some areas of the dance world, which promotes a belief that the body has to be pushed to its physical limits if an individual is to succeed as a dancer.[5] There are also links in Alex's drenched look with the post-coital sexuality of the previous dance sequence.

The various close-ups in this section provide a fascinating conceptualization of the dancing body. The spectator is not presented with a complete body, but one that is commodified into various fragments: the head shots show a 'driven body' with its obsessive will to succeed; the buttocks, as they writhe and undulate, represent a 'sexual body'; and the feet depict a 'sporting body' with its rigorous and regimented drill. The film therefore simultaneously sells a selection of corporeal constructions that effectively promotes and markets the dancing body through distinct but intersecting discourses. This type of scene, in which the protagonists rigorously rehearse, is seen across a number of 1980s dance films[6] and thus provides a popular image of the dancing body as a phenomenon that can only be achieved through time, effort, diligence and transformation.[7]

There are other elements of this sequence that also perpetuate the popular image of dance. For instance, at one point Alex sits with her legs spread wide apart while she vigorously bounces her upper back down towards the ground. Although the action bears similarities to a ballet-style stretch, this type of bouncing is considered to be damaging to the spine and is widely avoided in current dance practice. Likewise, she is later seen twisting, spinning and arching along the ballet barre in a rapid phrase of movement. Yet regardless of the fact that a barre is primarily used as a support for the slow, repetitive warm-up exercises, the authenticity of the dance class is not important as long as the popular-dance imagery is provided. A similar example is with Alex's leg warmers. Although dancers generally wear them pulled over their legs as a means of insulation, Alex wears them, throughout the film, pushed down below her calves as was popularized through the street fashion of the 1980s.

It is also worth examining the function of music in the film. In this particular sequence, the pop song 'Maniac' provides the accompanying sound for the dance. The lyrics, which state 'She's a maniac, dancing on the floor, and she's dancing like she's never danced before', back up the images of Alex's obsessive rehearsing, while the upbeat dynamic of the track acutely mirrors the vigorous and compulsive exercises. In addition to this narrative function, the song provides an important intertextual reference to the 'Maniac' single and the *Flashdance* sound-track that were released at about the same time to the film. The pro-motion of related merchandise is also an aspect of other 1980s dance films, such as the spin-off television series from *Fame* and the success-ful relaunch of pop group the Bee Gees who sang on the *Saturday Night Fever* soundtrack. This would then suggest that the dancing body within 1980s dance film is also a site of consumerism.

It is notable that the commodification of the body is operating in several ways. On one level this is purely financial, in that related merchandise, in the form of music, dance wear and exercise classes, generates more capital. On another level, the commodification of the 'woman's exercise body' serves a number of ideological purposes. Although this construction presents an empowering and independent image for women, it has been noted that the energy women invest in decorating and maintaining their bodies is a time-consuming regime (Featherstone, 1991). This is typical of popular culture, in which images are embedded with polysemic readings and thus allow for a variety of spectatorship positions (Fiske, 1989). Women may read Alex as a strong and independent role model, whereas men may see her as an erotic figure; or a working class reader may see her as representing a challenge to the cultural hegemony of high art when she wins her place at the elite, notably 'white', ballet school. It is perhaps because the cinematic institution is very much a commercial enterprise that *Flashdance* commodifies the body in numerous ways.

Within *Flashdance*, irrespective of authenticity, the popular concep-tions of dance are provided and maintained. The slim, youthful body, the ballet barre and rehearsal studio, the required attire of leotard and leg warmers, the virtuosic movement content, easily recognizable dance styles such as ballet, jazz and street dance, and the personal and economic struggle required to achieve success, all contribute to this perception. Much of this imagery derives from a hierarchical and stereotypical framework: the dancing body as sexually unleashed, bal-let placed at the pinnacle of aspiration, and the fetishized female body as a site of erotic spectacle. Although *Flashdance* readily employs the codes and conventions of the filmic apparatus in the form of close-ups, reaction shots and rapidly cut sequences, the familiar imagery con-structs a dancing body inscribed with popular discourses. The result is a conventionalized dancing body that not only facilitates polysemic readings (consequently making itself accessible to the widest possible audience) but is also commodified in such a way as to allow for com-mercial reward and dominant ideological positioning. The popular representation of the dancing body is coded to perpetuate certain economic, socio-cultural and ideological structures. Any element of choreographic and filmic innovation is overlooked in preference to popular images and mass appeal.

Television advertising and dancing bodies

Within the context of television advertising dance has been used to sell a range of products including building societies, mayonnaise, jeans, chocolate, soft drinks and watches. Whereas 1980s dance films implicitly promote social meanings through the dancing body, in television advertisements dance is explicitly used to sell specific commodities. This section therefore aims to investigate how the dancing body can be used to promote a product, how the televisual apparatus participates in this process and whether there is an ideological component to the way in which dance is employed. In order to do so I have drawn on two advertisements: the first is for a Cadbury's chocolate bar called Twirl and the second is for Hellmann's mayonnaise. The two advertisements rely upon different styles of dance and are therefore able to promote a distinct set of messages in relation to each product. Before looking at these examples in more detail, however, it is important to have some understanding of how advertising functions.

For the majority of people who live in a capitalist society, the practice of reading advertisements has become almost second nature. Individuals habitually participate in the structures of advertising without ever necessarily reflecting on the logic of this form. Yet advertisements are structured in a specific way and inscribed within the form are particular 'reading rules' (Goldman, 1992). There tends to be little differentiation between specific types of commodity in the sense that one washing powder is very much like any other. As the difference between commodity types is negligible, the aim of the advertisement is to locate a unique selling point (Williamson, 1978; Wernick, 1991). Thus the creators of advertisements connect certain images, objects, people, emotions or values to a product which then become associated or attached to it (Williamson, 1978; Falk, 1994). It is this formal technique of giving a product 'symbolic value' that characterizes all advertisements.

The images and ideas that advertisements draw upon derive from everyday life (Williamson, 1978). Advertisements extract 'meanings' from the 'real world' and recontextualize them in relation to a product. Hence meaning is transferred from one value system to another.

Goldman (1992) describes this as 'rerouting' meanings, in that the advertisement does not create new meanings, but rearranges existing ones. He suggests that this process implicitly creates 'false assumptions' through linking commodities with images and ideas beyond their immediate use-value. This process of appropriating visual signifiers from everyday life and attaching them to a product can be seen through a semiotic perspective. The 'signifier' is the product and the 'signified' is the meaningful image/association, and together they constitute the 'commodity-sign' (Williamson, 1978; Dyer, 1988). In other words, a particular image is linked to, and therefore becomes connected with, the product. This implies that buying the product will give the consumer access to the attributes associated with the image (Williamson, 1978), and it is for this reason that advertising is said to create 'false values'.

Any analysis of the way in which advertising operates is inadequate without some reference to the 'reader' who is integral to the interpretive process. Although it is the formal logic of the advertisement that links a product and an image together, it is the reader who must construct a meaningful connection between the two (Williamson, 1978; Dyer, 1988). The way in which a reader is able to make sense of an advertisement is reliant on two central factors. First, as advertisements appropriate images and ideas from the reader's everyday experiences, potential meanings are already in circulation. Second, advertising is highly intertextual and 'reading rules' are located in the frameworks of other advertisements. That is to say, we inhabit a culture of advertising imagery and we as consumers are familiar with its discourses. Yet, although advertisers aim for 'preferred readings', it is clearly not possible to guarantee them, as readers are neither homogeneous nor passive (Goldman, 1992). Consequently, readers may purposely, or unwittingly, 'misread' advertisements. Advertisements therefore lay themselves open to interpretation.

In order to demonstrate how this theory works in practice it would be useful to turn to the two examples. The first advertisement under analysis is for Twirl, a chocolate bar produced by the confectioner, Cadbury. The advertisement begins with a rapid series of shots, focusing on a group of dancers, dressed in a variety of fashionable black and white practice clothes, rehearsing a short jazz routine. The movement vocabulary includes the regular repetition of a multiple spinning action accompanied by an upbeat jazz melody. As the dancers reach the end of the sequence a male 'rehearsal director' enters saying, 'Okay, take five' and a female voice sings along to the sound track, 'Take five, just take five'. Several shots ensue, including a number of close-ups of the product and of the people who sit around chatting, laughing and eating Twirl bars. A male 'voice-over' then comes across the sound track and states, 'Cadbury's Twirl. Two folded fingers coated in delicious Cadbury's chocolate. Take five for a Cadbury's Twirl'. The big close-up images that accompany the speaking voice depict a pool of chocolate, rippled chocolate layers, and the 'chocolate finger' being coated. Towards the end of this speech the group begins to rehearse again. The penultimate long shot is of the whole group spinning, or 'twirling', in silhouette and the final shot shows the words, 'Take five for a ... ', with a Twirl bar placed diagonally underneath the 'copy'[8] to complete the phrase.

In this instance, the commodity-sign is constructed through the connection between a chocolate bar and dancing bodies. The advertisement does not imply that the consumer can become a 'dancing body', but rather that the consumer can gain access (through the chocolate bar) to what the dancing body represents. With this particular example, the dancing demonstrates youth, energy, athleticism, sexuality, commitment, motivation, companionship, potential stardom, teamwork balanced with individuality, and a slim, strong, attractive and healthy body. Images of youth, beauty, fitness and so on are a recurring theme in advertising (Featherstone, 1991). Yet aside from this connection, the reader is further inserted into the logic of the advertisement through a number of visual–verbal puns.

The word 'twirl' is not only the name of the product, but also describes a dance step, which is repeatedly presented in the image. It is significant that the term 'twirl' is not part of a specialist dance vocabulary but is generically used to describe any kind of spinning action. This use of popular terminology makes the play on words accessible to a wider range of readers than if the advertisement had used a specialist vocabulary. The use of punning continues, although perhaps less explicitly, with the phrase 'take five', which is requested by the 'rehearsal director', and then reiterated in the sound track,[9] the image, and the final copy. Within a variety of 'performance' discourses the term 'take five' can be read as a sign to take a break, which the dancers literally do. Yet on another level, the notion of 'take five' is a command for consumers to take a Twirl as part of their break. The use of imperative tense and the numerical quantity implores the reader not only to consume once, but to consume many times. Williamson (1978) notes that puns, puzzles and jokes are a regular feature of the hermeneutic framework of advertising. She suggests that advertisements create a symbolic world that has to be deciphered. Through requiring the reader to solve various puzzles, the advertisement deflects the reader's attention from the real interpretive task of making a connection between two unrelated phenomena, that is the signifier (product) and its signified (image).

Of course, it may also be possible to make a non-preferred reading. Instead of linking the chocolate bar to a dancing body that is read as attractive and virtuosic, the reader could make a connection between dancing bodies and a lack of time to eat proper meals, the need for immediate energy from sugar snacks and comfort food as a means to deal with the emotional imbalances that coincide with this uncertain career path. These ideas carry implicit notions of eating disorders and it is pertinent that dance has been linked with other chocolate products in television advertising.[10]

As an end to this analysis it may be useful to consider briefly some of the formal televisual devices that are employed in the Twirl advertisement. The dance sequences are subject to a rapid edit, which jumps from long shots, to close-ups, medium shots and top shots. This play on camera work and editing bears similarity to the rapid filming style of the 1980s dance film. In many ways the actual shots are highly conventional. The close-ups either depict the performers' faces, in order to reiterate their commitment and attractiveness or, in other instances, focus on gyrating hips, an act that locates the dancing body within overtly sexual discourses. The long shots are used to display the overall choreographic pattern and the top shots are employed as part of the vertiginous editing, which rapidly jumps from one image to another. The effect is that the image becomes a ceaseless spectacle of virtuosic bodies while the rhythm of the edit complements the upbeat jazz score. Yet although the edit is somewhat 'dizzying', the images carry a discernible narrative: there is a degree of characterization in the 'performers' and their 'rehearsal director', a coherent spatial logic in that the scenario takes place in a rehearsal studio, and a sense of linear temporality as the dancers rehearse, take a break, and then go on to rehearse/perform the sequence again. Both in terms of dance content and formal style, the advertisement makes intertextual references to the 1980s dance film and it appears that such references are no accident.[11] The allusions to 1980s dance film not only provide a popular representation of dance but also link the Twirl bar to the various discourses inscribed in the 1980s dance film body. McRobbie (1990) asserts that these films promote personal success, romantic fulfilment and 'fantasies of achievement', which in turn can be rerouted back into the Twirl commodity-sign. Thus it would seem that advertising is far from innocent and clearly plays an ideological role.

The second advertisement under consideration is for Hellmann's mayonnaise. It is significant that this advertisement also connects dancing bodies to a food type, but in this instance the dancing body is used to carry a completely different set of messages. The advertisement is set to Tchaikovsky's *Nutcracker Suite*, and brightly coloured images are placed against a stark white background. The majority of the advertisement alternates between medium shots of dancers dressed in either unitards or flowing ballet-style dresses and close-up images of fresh salmon, sprigs of tumbling lettuce, carrots and courgettes being peeled or sliced, and shots of Hellmann's mayonnaise. Meanwhile, a male voice-over reads the copy that appears on the screen: 'Without a choreographer there is no ballet... Without Hellmann's there is no salad'. The final image depicts two dancers in 'adage poses' who are then transformed into jars of Hellmann's and Hellmann's Light. Both the copy and the voice-over state, 'The only mayonnaise'. The only other copy is at the beginning of the advertisement, which is written in extremely small print at the base of the image, and states, 'Can help slimming or weight control as part of a calorie controlled diet'.

In this example, the commodity-sign connects mayonnaise with popular images of balletic bodies: flowing costumes, ballet music, pointe work, pirouettes and arabesques. On first glance it may appear that the point of connection is in the message, which questions where the ballet and salad would be without the choreographer and mayonnaise respectively. The assumption is that the former could not exist without the latter. Yet it could be argued that this is simply a red herring and that the actual link is between the Hellmann's mayonnaise and the weightless, ethereal aesthetic of the 'balletic body'. The associations of mayonnaise with salads (and general 'healthy eating') are subtly, yet persuasively, connected to the small print, which draws reference to potential weight loss.

This idea is further highlighted on closer examination of other formal properties of the advertisement. Through various means, the dancers' bodies both literally and metaphorically represent the different food types. First, the dancers resemble the salad and mayonnaise in appearance, in terms of costume style and colour. For instance, two women wear bright orange unitards with green scarves to allude to the carrots and another two wear blue or red dresses, with bright yellow tights, to match the colour codings of the two jars of mayonnaise.[12] Second, the links between the dancers' bodies and the food types are suggested in the edit. For example, the two dancers dressed in orange directly precede the carrot close-up, a top shot of a dancer spinning in a red dress precedes the image of the red lid of the Hellmann's jar being unscrewed, and so on. The movement vocabulary is similarly reminiscent: when a courgette is sliced down the centre, two dancers fall apart in a lunge, and as lettuce leaves tumble downwards a dancer is depicted in the descent of an aerial step. The connection between the ballet body and low-calorie foods, however, goes much further than surface appearance. As with the food types that make up the salad, the dancers are literally 'healthy', light in weight and physically constituted through a low amount of calories. They are the embodiment of the link between ballet and low-calorie foods. Once again, however, several non-preferred readings can also be made. One reading derives from the stereotypical imagery inscribed in this advertisement. The associations of dancers with diet, weight loss and the slim, ideal body type easily lead to the concept of anorexia and its related psychological problems of poor self-esteem and negative body image. An alternative, sceptical reading may take note of the irony in linking mayonnaise to salads and images of weightlessness, when mayonnaise is highly calorific in actuality.

As with the 1980s dance film, the dancing bodies within the two advertisements are similarly inscribed in conventional dance discourses: the movement vocabulary is derived from established dance techniques; traditional practice or performance clothes are worn; and certain 'live performance' codes are written into the image, such as the empty studio/stage settings and the way in which the dancers 'project' to the camera. The camera work and editing devices allude to other popular representations of dance in film and television. Consequently it is through the recognition of this popular imagery that the reader is able to connect certain ideas and values to a particular product and make either preferred or non-preferred readings. In these two case studies, at least, the preferred readings clearly play an ideological role in that the advertisements trade in stereotypical imagery and dominant values.

Dance and pop music video: a musicology of the image

Another screen genre through which the dancing body may be seen as a vehicle for promotion is the music videos that accompany many popular music singles. Since its inception in the early 1980s[13] it has become apparent that the pop music video clip is less a commodity form than a promotional one. Music videos were originally intended to promote singles. Although the videos are relatively expensive to produce, they are effectively paid for through increased record sales and, to a lesser extent, through cable and satellite subscriptions to music channels, such as MTV and VH1, and home video rentals (Goodwin, 1993). Frith (1988) notes that music television initially prompted an acceleration of record sales, as the market had previously been reliant on the slower processes of LP and tour promotion. There has since been some decrease in the promotional effectivity of music video in terms of record sales, but in place of this music television has created a wider marketing network. Goodwin (1993) suggests that the music television audience is now the 'product' that appeals to advertisers, while Frith (1988) describes the way in which record companies charge television stations for the rights to play their videos. Indeed, a whole consumer culture has evolved through the intertextual promotion of items such as jeans, beer and cars that appear in the music videos themselves (Goodwin, 1993).

Although dance does not feature in all music videos, popular dance forms are often closely linked with current music trends. Consequently dancing bodies frequently operate as a major component of music videos. The dance element may be performed by audience members, professional dancers, or the singers and musicians of the band, and can range from loose, improvised movement through to tightly choreographed routines (Buckland, 1993). For instance, the video for *Fly Away* (1999) by Lenny Kravitz is based on a live gig and shots of the band are intercut with images of the crowd. The audience members appear to improvise to the music and the movement ranges from general jumping and waving through to more specific articulations of the limbs and torso. In contrast, Will Smith's *Wild Wild West* (1999) features a chorus of backing dancers who perform slickly executed routines. The video culminates in a theatrical show number: situated behind Smith in a tight unison formation, the dancers perform stylized movement phrases based on Wild West imagery such as riding and lassoing.

Dance in music video can equally be used as an indication of the recording artist's skill (Buckland, 1993). Singers such as Madonna and Michael Jackson, who paved the way for dance in music video during the 1980s, are known for their use of complex and virtuosic routines evident in videos such as the former's *Vogue* (1990) and the latter's *Thriller* (1982). The legacy of these artists is perhaps seen in the recent surge of boy and girl bands, such as The Spice Girls, Steps, S Club 7 and Another Level, for whom staged dance sequences are an essential component of both their performance and video work (Dodds, 1999). Significantly, although other performers may lack formal dance training or technical expertise, dance can still be a key component of their musical identity. One such artist is Jay Kay, the lead singer from Jamiroquai, who employs a distinctive and personalized movement vocabulary. In the video for *Canned Heat* (1999), as well as leaping through walls, diving into a television screen and swinging from a light fitting, he pulses in a knock-kneed stance and executes nifty footwork and snaky gesticulation to match the funky soundtrack.

In some instances the style of dance may correspond to the musical genre (Buckland, 1993). Ricky Martin's *Livin' La Vida Loca* (1999) draws on a Latin American trend both in the content of the song and the style of the music. This theme runs through into the video as his backing dancers perform a generic (and somewhat stereotypical) Latin American movement vocabulary of flicking heads, sharp turns, pelvic isolations and fast footwork. Another example is the video for *It's Like That* (1997) by Run DMC vs Jason Nevins. The single is a reworked version of a track by seminal 1980s hip hop band Run DMC and the video is of a 'dance out' between two rival gangs. To complement the musical lineage, the video draws on classic 1980s street dance forms, such as breakdancing, body popping and locking. The way in which the televisual apparatus is employed in music video is paramount to the sense of motion. Buckland (1993) asserts that the energy of the camera is intended to parallel the dynamic of the performers. In the video for *Let Me Entertain You* (1997) by Robbie Williams, the dizzying camera work and rapid edit produce a relentless torrent of movement images as Williams twitches manically, postures provocatively and leaps wildly across the stage. The energy of the filming style gives the viewer a greater sense of participation, as if she or he might be seeing events from the performer's perspective (Buckland, 1993).

Before looking at two specific examples in detail, it would be useful to examine the work of Goodwin (1993), whose theories form the basis of an analytic framework for the study of music video. Goodwin (1993) suggests that in some of the early scholarly writing on music video the aural element is marginalized in favour of the visual.[14] Yet he stresses that the images of music video are designed to illustrate the sound (rather than the other way around), and both musicians and audiences habitually visualize music during its production and reception. To avoid a purely visual bias, Goodwin (1993) has conceptualized a 'musicology of the image' as a methodology for the analysis of music video. He argues that, in reference to music video, it is essential to take into account the relationship between the music and the image. He uses the concept of 'synaesthesia' to inform this theory, which he defines as 'the intrapersonal process whereby sensory impressions are carried over from one sense to another, for instance, when one pictures sounds in one's "mind's eye"'(p. 50). This notion is pertinent to music video in which images are constructed through the visual associations of a soundtrack.

Goodwin (1993) states that music is made up of various components, such as tempo, rhythm, arrangement, harmonic development, acoustic space and lyrics, which he then identifies in relation to the images of music video. For instance, the tempo of the music may be reflected in the speed of the camera movement, the style of edit, the events in the video and through various special effects. Likewise, the rhythm may be echoed in the 'pulse' of images. He suggests that the arrangement of a song is dependent on the relationship between a number of factors such as voice, rhythm and backing, and this can be carried through into the choice of images. For example, the voice can be foregrounded through close-ups of the singer's face and key rhythmic moments are focused on through shots of the musicians. Similarly, the video may also mirror key shifts in melody, while the content of the lyrics can be illustrated, either literally or metaphorically, through the choice of image.[15]

Many of these ideas are supported by Frith (1988), who suggests that the aim of music video is to create the impact of live performance on the small screen. He suggests that 'movement' becomes a metaphor for sound, through such devices as fast cutting, swirling bodies and visual excess. He notes, 'we are overwhelmed with images to compensate for the essential feebleness of TV sound' (p. 216). That is not to suggest, however, that all music video is permanently tied to the form and content of a song. Although visualization can illustrate the meanings of a song, it can also amplify or contradict them (Goodwin, 1993). For instance, a music video can use images to create layers of meaning that would not be apparent from just the lyrics, or it can be used to promote commodities other than the song. To an extent music video is intertextual, but it also contains a certain degree of logic and coherence.

In order to examine constructions of the dancing body within music video I want to look at the work of Philippe Decouflé. Decouflé is a particularly interesting example because he is both a choreographer within the field of art dance and a director of commercial dance on screen. The similarities between his choreographic work in theatre and the representations of popular video imagery are clearly identified by Bozzini (1991), who remarks:

> Inspired by comic strips and film clips, this superb dancer is one of the few in France to epitomize to the very tips of his fingers the young man about town who-knows-how-to-make-an-ad and who can create dance which is so much like advertising you can't tell one from the other. (p. 38)

It is perhaps not surprising that Decouflé's work in theatre dance bears a remarkable likeness to the images of advertising and music video. In addition to his stage work he has directed music videos for New Order (*True Faith*, 1987) and Fine Young Cannibals (*She Drives Me Crazy*, 1988), and directed television advertisements for Gervais (1988), Dior (1989) and Polaroïd (1989). The following analysis therefore focuses on the music videos for *True Faith* and *She Drives Me Crazy* and employs Goodwin's (1993) musicology of the image in order to address the links between the soundtrack and the visual imagery.

The first example is *True Faith*, by the band New Order, which Decouflé choreographed and directed in 1987. The video consists of various surreal 'dancers' who engage in strange and unfathomable behaviour, alongside images of the band. The piece begins with two figures, dressed in brightly coloured rubber 'dungarees', who slap each other in time to the drum beat. As the camera pulls back, a large grey foot comes into view. This belongs to a dancer, wearing a long red 'dress' enmeshed in a metal spiral, who sits on a platform above the action and appears to have only one leg. He wears a white bib and a metal frame attached to his head with a small screen placed directly in front of his eyes. As he reaches up to twist a knob on the frame, the image on his screen cuts to a close-up of the lead singer, Bernard Sumner.

The use of close-up to 'frame' Sumner is a typical convention in both music video and pop photography. It tends to be employed as a promotional device as it provides a 'face' for the band, both in the sense of the actual singer (who is usually the integral front man or woman) and the band's 'image', in terms of its style and genre. Throughout the video there are regular images of Sumner, the keyboard player and the drummer. The pop iconography inscribed in the image implies that the band is in the context of a live performance: the musicians are surrounded by darkness, except for occasional flickers of bright white light, which suggest the flash of a camera; Sumner uses a microphone and has a guitar slung on his back; the camera is placed in a low angle position as if to focus up to him on stage; and a rear view shot depicts an audience who dance along in time to the music. Whether it is an actual 'gig' is of no importance, as long as the codes and conventions of live performance are written into the image.

Although many of the images consist of the surreal collection of 'dancers', the video is coded with familiar pop iconography. The 'dancers' are situated in striking visual imagery and televisual devices are used to manipulate and distort their movement. For example, a shot of three dancers, dressed in striped 'Oskar Schlemmer-inspired' costumes, is played in reverse so that they appear to run and leap backwards. Later they are seen bouncing on a mattress and the image is both speeded up and slowed down. Yet, in spite of the technological play on the dancing body, many of these images relate back to the rhythm and tempo of the music. For instance, the two men who slap each other during the introduction do so in time to the drum beat and the characters who bounce up and down follow the rhythm of the melody precisely. The use of technology to affect movement reflects the technological persona of the band which was known in the 1980s for its use of synthesized music. The technologically mediated bodies both complement and reiterate the high-tech image of the band.

It could be argued that there is an explicit challenge to realist devices in *True Faith*, in terms of costume, movement and action. For instance, a woman, encased in a black rubber shell (which is attached to a spring so that she rebounds back and forth), executes a complex, but incoherent gestural sign language. The 'one-legged man' unsuccessfully attempts to stack three orange objects, in the form of a sphere, a cube and a tetrahedron, while the dancers in rubber dungarees and the 'Schlemmer' costumes begin to wrestle among themselves. Each character appears obsessed by impossible and irrational tasks. Yet regardless of these seemingly 'incoherent signifiers' there is an element of logic to *True Faith*.

Although the images do not directly correlate to the lyrics, the singer recounts a sense of inner turmoil and confusion through the words of the song. For instance the chorus goes:

> I used to think that the day would never come
> I see the light in the shade of the morning sun
> I know the sun is the drug that brings me here
> To the child that I lost, replaced with fear.

It could be suggested that the bizarre and incoherent images sum up the sense of confusion implied through the lyrics. There is also another level of logic to the images of *True Faith* in that they constantly refer back to various codes and conventions of the popular music context: framing devices are employed to promote the lead singer as a 'face' for the band; the video is encoded with 'pop concert' imagery; and the dancers' actions parallel musical components of the song, such as rhythm and tempo. The innovative and surreal visuals also promote a particular image for the band. Buckland (1993) argues that the choice of Decouflé, as an artist situated in French avant-garde dance, and his particular use of movement imagery gives the band the status of serious musicians. There is clearly a certain 'high art' intellectualism implicit within the references to surrealism, Bauhaus-inspired costumes, and avant-garde dance. Hence, it could be suggested that, when the music context of the video is taken into account, these apparently incoherent and open-ended signifiers are not as illogical as they might at first seem.

The following year Decouflé also choreographed and directed *She Drives Me Crazy* (1988) for the band Fine Young Cannibals and, stylistically, the video is similar to *True Faith*. Once again, shots of the band are intercut with images of dancers, dressed in zany costumes, who obsessively engage in irrational and impossible tasks. As with *True Faith*, many striking visual devices are also employed in *She Drives Me Crazy*. For instance, a man in a black-and-white striped suit is framed in a tight, full shot with his knees bent outwards and his head creased sideways. He swings his arms from side to side, while his knees twist in and out, and the overall impression is that he is too tall to fit into the tight frame, hence the cramped-looking movement. Another example is of a man dressed in a yellow suit, trousers rolled up at the knees, who stands on his hands over a perspex floor. The camera is placed directly beneath the perspex, thus the viewer sees his body from an extremely peculiar angle. There is also an absence of realist strategies: a man, whose body is covered in large, red-satin cushions, springs across the floor and repeatedly bounces onto his stomach and back; and a top shot depicts two men, dressed in yellow and black suits, who have a jumping race for no apparent reason.

Yet once again, although the images are absurd, many of them tie in with musical components of the song or general pop music conventions. Similarly to *True Faith* the actions of the dancers are used to reflect key musical structures. For instance, the men in yellow and black suits execute their jumping race in the same rhythm as the introduction to the song and the camera disjointedly zooms in and out towards the lead singer, Roland Gift, so that he appears to whizz towards and away from the screen, also to this rhythm. In addition to this there are other pop music conventions that come into play. During the vocal sections there are regular close-ups of Gift. Unlike the 'poor-quality' image of Bernard Sumner in the *True Faith* video, who is masked by his microphone, shot from an unflattering low angle and depicted through the blurry focus of live performance, Gift is positioned face on to the camera, against a number of brightly coloured backdrops, which highlights his good looks and 'pin-up' status. Just as a backing singer comes in on the soundtrack, two painted faces pop out of brightly striped cylinders to mimic the vocals and, during an instrumental section, the 'black and yellow men' dance against a black and white split screen backdrop. Although their movements show a stylized and distorted type of disco dancing, the device of using dance sequences during instrumental sections is a regular pop music convention. The video is clearly inscribed with established pop music practices, from the use of backing singers and dancers to the star 'pin-up'. The lyrics to *She Drives Me Crazy* are about the way in which the 'singer' fails to comprehend the actions of 'a woman' and, as with *True Faith*, the irrational images could be said to reflect his confusion: 'I can't stop the way I feel. The things you do are so unreal...She drives me crazy. Like no one else'.

It could be argued that the 'arresting images' of the two music video examples are intended to encourage repeated viewings. Although both employ innovative or experimental imagery in parts, any aesthetic radicality is counteracted by the close links between the visual components and pop music conventions. Whether it is through associating the cutting-edge imagery with music practices such as pin-up singers, backing dancers and formal components of the song, or by giving the band a zany or arty image, the videos serve a promotional function. In a number of ways the imagery either advertises the single or the band. In this respect it appears that images of the dancing body within pop music video share a similar terrain to the commercial representations of 1980s dance films and television advertising. Artistic innovation is underplayed in favour of existing conventions and commercial appeal.

The translation of theatre dance to screen

So far this chapter has focused on works that are situated at the commercial end of screen dance. This includes dance that is created for a mass audience and dance that serves a promotional function. The remainder of the chapter therefore turns to art dance on screen. It is suggested in Chapter 1 that the characteristics of a live body do not always effectively translate to a screen body, and the capacities of a screen body cannot always be duplicated by a live body. This is clearly problematic for television adaptations of stage choreography in that theatre dance is designed in relation to a live body that has certain spatial, temporal and dynamic capabilities. As theatre dance is conceived for the stage, it is constructed with the codes and conventions of that context in mind. Consequently, television adaptations of stage dance must negotiate the disparities between an already existing dance work and the televisual form. I therefore want to consider how screen adaptations employ the televisual apparatus in order to create a faithful translation of a stage performance. With this analysis I am less concerned with explicit dance content and subject matter than with how televisual devices can be used to 're-present' an image of the 'stage body'.

Before examining this in more detail, it is necessary to clarify the parameters that characterize the translation of theatre dance to screen within the context of this section. It could be suggested that television adaptations of stage choreography can be divided into two categories: those that attempt to make a 'faithful translation' of a stage work to screen; and those that rework the original choreography specifically for the screen in order to create a new work in its own right. For the purposes of this section, I am particularly interested in the first category of television adaptation and it would be useful to elucidate what might be meant by the concept of a 'faithful translation'.

The televisual apparatus unavoidably distorts and manipulates the body and, consequently, a televisual replication of a live, dancing body simply cannot exist. Yet some stage pieces are recorded in order to preserve and document the work of a particular choreographer, often with the intention of bringing it to a wider audience. Hence such 'television adaptations' are intended as a faithful record of an existing stage work, and the aim is to retain a clear sense of the original choreography. Accordingly, the director attempts to capture the 'essential components' and choreographic design of the original, while also observing certain features of the televisual form; an unedited, long shot is clearly inappropriate for the medium. As perception is such a variable phenomenon it is perhaps worth qualifying the notion of a 'faithful translation' by stating that it aims to document faithfully an existing stage work in the view of a particular director. This then acknowledges that multiple 'realizations' may exist. With this is mind, the remainder of this section seeks to address some of the difficulties of translating stage work to screen and to suggest how a faithful version of the original might be produced.

For the first few decades of television the majority of dance programming was devoted to ballet, in the form of full-length works, excerpts, interviews and documentaries (Rowson Davis, 1982–83; Penman, 1987). In the case of full-length performances, there has been much debate as to the most appropriate approach to filming ballet (Brooks, 1987–88; Mackrell, 1993; Meisner, 1984). The television adaptation of ballet is problematic as the classical form and aesthetic are notably incompatible with the television medium (Lockyer, 1993). The large *corps de ballet* requires a long shot to capture all of the dancers on screen but results in screen figures that appear minute and distant. In contrast, when the camera focuses on medium or close-up shots of the principal characters, the composition of the accompanying *corps de ballet* is then lost. The close-up shot can potentially ruin the ethereal aesthetic of the ballerina; a close-range view may reveal a sweating, muscular body and, as a result, shatter the illusion of effortlessness. So whereas continuous long shots make dull viewing for the spectator, close-up perspectives are in danger of distorting the classical aesthetic. Due to the large number of dancers and the elaborate set designs that generally accompany full-scale ballets, it is more cost effective to record these works during live performances. This, however, sets up various limitations in terms of camera position and the number of possible 'takes'.[16]

The genres of modern and postmodern dance, which are informed by a range of contemporary dance techniques, are considerably less problematic to film than ballet. Although there are exceptions, 'contemporary dance' tends to require fewer dancers, it rarely employs *grand tableau* or spectacle, and its aesthetic does not demand an illusion of effortless ethereality. Hence, close-up shots of muscular and labouring bodies do not necessarily present a problem. Due to the smaller scale of most contemporary dance works, it is possible to record them in a studio setting, which allows for greater flexibility in terms of rehearsal time, camera position and potential error. As Graham-based contemporary dance arrived in Britain only during the 1960s and took several years to take root and establish itself, it was not until the late 1970s and early 1980s that it began to be screened on television. This period saw the development of a long-standing relationship between television director Bob Lockyer and the 'Graham trained' choreographer Robert Cohan, who was the founding artistic director of London Contemporary Dance Theatre (LCDT).

Lockyer filmed a series of LCDT works intended to be 'faithful translations' of stage choreographies to the television screen. I therefore want to investigate the ways in which some of the 'space–time–energy limitations' are resolved when theatre dance is translated to screen; how the televisual apparatus is utilized in order to create a 'faithful rendition' of the original; and what type of dancing body is constructed through these codes and conventions. The focus of this analysis is primarily in relation to the Lockyer/Cohan production of *Cell* (1969, TV 1983), but, in order to examine whether any alternative approaches or paradigmatic trends come to the fore, there are some references to other Lockyer/Cohan collaborations. These include *Stabat Mater* (1975, TV 1979), *Waterless Method of Swimming Instruction* (1974, TV 1980), *Forest* (1977, TV 1980), and *Nympheas* (1976, TV 1983).[17] *Cell* was choreographed by Cohan and employs socio-psychological subject matter as its theme. It is made up of six dancers, who are divided into three male and female couples, and takes place inside three white rectangles, which suggest the walls of a cell. The piece is divided into four sections that deal with various ways in which humans interact within the modern world. This analysis examines the first section.

The film opens with an establishing long shot depicting the empty cell. As the dancers enter one by one from the rear wings and fall to the floor, the camera slowly begins to zoom in, but retains all of the bodies in full view. The choice of shot is conventional in that it creates a clear sense of location: the spectator is given a complete view of the set and is able to see all the action from a stage front perspective. In fact, the majority of the dance is filmed from a frontal or diagonal camera position in relation to the back wall of the cell, which clearly mirrors the viewing context of a stage audience. As the last dancer falls to the floor the film cuts to a close-up, which captures the bodies scrambling towards the walls. The particular angle of this shot is a very effective use of screen space in that it is filled with several pairs of legs. Often with a group of dancing bodies on television, the upper part of the screen is used only to depict the performers' heads or the space above them (Lockyer, 1983). In this instance, however, as the camera is low to the ground in a diagonal mid-shot and the dancers are lying down, it is cleverly able to focus on several pairs of legs at the top, bottom and sides of the screen space. The camera then zooms back to a long shot, which once again positions the spectator as if in a theatre context by showing a view of the whole group and how each dancer is placed in relation to the others.

The camera cuts to a series of medium or full shots of the various duets that ensue. For most of the time the camera is static and focuses on 'complete bodies' within a tight frame. Any camera movement that does take place is slow and sustained. Occasionally it cuts to a medium or close-up shot to examine a duet in more detail, but it then cuts back to a full shot to reveal how the two bodies are moving in relation to the rest of the group. This device of cutting back and forth allows the spectator to see some element of detail while retaining an overall sense of the choreographic pattern. Indeed, *Cell* is a compatible piece of choreography for the television medium. When the group is not moving in unison, the focal point of action is always very directed; for example, during the many duets that take place, the remaining dancers are almost always in a static position, which allows the camera to focus solely on the duet without fear of missing any other action.

Many of the shots are filmed so as to give a strong impression of foreground and background, which prevents the choreography from appearing flattened by the two-dimensional screen. In several instances a shot is taken from a sharp diagonal that is close to the action, so that the bodies at the front of the frame appear enlarged while the bodies at the rear of the frame appear minute and distant. In some ways these extreme perspectives reflect the content of the piece; the distorted and slightly askew camera angles emphasize the social and psychological breakdown that forms the subject of *Cell*. For example, in one shot, a man is shown scrambling towards a pair of legs that keep stepping backwards. Compositionally, he and the pair of legs appear enlarged in the foreground, while several other pairs of legs appear dotted around in the background. The close-up camera work, bizarre angle and distorted sizes effectively magnify his moment of subservience and panic. Yet although the image is unconventional in terms of stage choreography, because this is clarified in the previous and following shots the viewer has a clear idea of how the man's body is situated in relation to the rest of the bodies. This specific shot is used more as a way to mirror the psychological content than as an independent play of the televisual apparatus. Unusual camera angles are far less common in *Waterless Method of Swimming Instruction* or *Forest*; the camera is positioned so that all of the bodies are evenly distributed in space.

There are also instances in which the content of the work is reflected through formal means. For example, a number of close-ups take place that magnify facial expressions and thus reflect the psychological nature of *Cell*. This device is echoed in the close-ups of the 'mother figure' in *Stabat Mater* and in the comical 'sunbather' in *Waterless Method of Swimming Instruction*, both of whom are set up as 'characters', or recognisable figures, with whom the viewer can identify when they reappear throughout the pieces. As the subject matter of *Cell, Stabat Mater* and *Waterless Method of Swimming Instruction* is 'human affairs', the close-up is an appropriate vehicle through which to draw attention to facial expressions, which can, in turn, reveal emotional resonances and individual details. In contrast, *Nympheas* is a far more abstract work and, as a result, the majority of the dance is filmed using long and full shots.

Although the speed of the first section of *Cell* is relatively slow paced, the dynamic behind the movement is sharp, direct and attacking, with some use of suspension. In several instances the camera work and editing capture this well. For example, at one moment there is a medium profile shot of a man kneeling down and a woman by his far side with her back to the camera. Just as she begins to fall slowly backwards the camera suddenly cuts to a (stage) front view so that the man is now face on to the camera as the woman falls in profile into his cradled arms. The specific camera positions and cut that are used for this fall effectively capture the sense of energy behind the movement. If the woman had fallen 'back-on' to the camera, then the spectator would have had less sense of her travelling through space. The sudden cut from her falling back to falling in profile, however, neatly mirrors the sharp dynamic of the fall. When a female dancer runs frantically back and forth among the others, the camera again contributes to the movement dynamic. It moves along with her so that all that is seen of the other dancers is odd limbs and torsos that briefly come into view. The camera movement creates another layer of kinesis. The series of cuts is also more rapid than previously, which suddenly encapsulates the speed and jerkiness of the dancer's zigzagging pathway.

As this section draws to a close the couples begin to move in unison and the camera work returns to a series of long shots so that the spectator is able to see the overall choreographic pattern. As each dancer exits at the end of the scene she or he is shown in a medium shot, and when the final dancer exits the camera pulls back to reveal the dark studio surround of the 'cell' as in the beginning, which closes the scene as if from a stage perspective. In fact, in all of Lockyer's work for LCDT there is an aim to minimize the spectator's awareness of the televisual apparatus at work (Lockyer, 1993): the camera movements are slow and subtle, the shots are not subject to a rapid amount of cuts and there are no unusual perspectives that completely distort the choreography.[18]

This analysis would suggest that a particular representation of a dancing body is constructed in the transition of theatre dance to screen. Although there is a selection of close-up and mid-shots in *Cell*, the majority of shots reflect a 'complete body'. The televisual apparatus is employed in such a way that the screen body is constructed as an 'illusion' of the live body, through the attempts to achieve spatial, temporal and dynamic consistencies with the stage performance; the work is predominantly filmed from a stage front perspective; specific camera positions construct a strong impression of depth in the screen; and the slow and sustained camera work reflects the steady pace of the piece, though the occasional cut or camera movement effectively captures a moment of dynamic suspension or attack. The dancing body is clearly inscribed with the codes and conventions of stage choreography. To some extent the spectator sees a number of constructions of the dancing body that could not exist outside the television medium, through devices such as close-up shots and edit points; yet the television adaptation of *Cell* operates as an illusory reconstruction of the stage context, reflected in the complete bodies, the cohesive choreographic patterns, and the 'stage front' perspectives. Although the television version of *Cell* constructs a screen body, it is very much 'framed' as a stage body. There is no sense of experimentation between the dance and the televisual apparatus; instead the work is a 'faithful document' of the original.

Early dance for the camera

The final case study continues to examine art dance, but from the perspective of dance created specifically for the television medium. It focuses on two early examples of dance choreographed for television in order to address the scope of creative investigation that took place at this time. A study of this early work provides the groundwork for Chapters 3 and 4, which address contemporary examples of dance conceived for the camera. This data can act as a useful gauge to mark the extent to which ideas and approaches to filming dance have evolved. The works under investigation are two 'ballets' originally created for television: the first is *Contrasts* (1956) and the second *Houseparty* (1964), which forms the main focus of the analysis. These are two of the earliest examples of dance works commissioned for television that are still in existence.[19] This section therefore sets out to examine how the televisual devices are employed, to analyse the movement content and to consider the type of dancing body that is constructed in early examples of 'dance for the camera'.

The television ballet *Contrasts* was screened as part of a longer programme entitled *We are your Servants* (1956). The programme was designed as an anniversary celebration of the television industry, ten years on from its post Second World War 'reopening' in 1946. *We are your Servants* is made up of many short segments that include anecdotes from presenters, celebrity interviews, and 'star turns' by singers, comedians, actors, dancers and an orchestra. The programme opens with three dancers, a male and two females, who 'warm up' together to the jaunty sound of a piano. It is clear that, by this era, television had got to grips with a number of editing devices, one of which was the possibility of utilizing several geographical locations within a short space of time. In the ensuing shots the dancers are revealed hopping and skipping through BBC corridors, the wardrobe department, the staff canteen and various make-up rooms. Within this 1956 film the viewer not only saw new spatial possibilities for dance but also witnessed the presentation of dance within unconventional contexts. As an end to this sequence, the dancers leap through a paper window in their practice clothes, but then emerge on the other side wearing tuxedos and ball gowns. The potential to employ editing as a play on temporality had obviously come into usage by this time.

Yet regardless of this use of the 'cut' as a means to manipulate time and space, the remainder of the dance pieces are filmed with a comparatively uninventive approach. This may be due to the fact that the first piece is somewhat comical and is accordingly matched by the playful choice of shots, whereas the following pieces are intended to be taken more seriously. For instance, the ballroom dancing couple, 'Boyer and Ravel' who appear a little later, are filmed from a single, unedited long shot. Similarly, the 'Television Toppers', a line of Tiller girls, are predominantly filmed from a face-on, long shot. Ten or so women are tightly squeezed into the frame and only two or three cuts take place, one of which shows the women from a diagonal perspective. It is possible that these two sequences existed prior to the programme; there is no suggestion that they had been specially designed for the programme, and ballroom dancing and Tiller girl sequences are established forms that derive from contexts outside the television medium (that is the social dance scene and 'variety' stage shows). The final piece of dance, however, was commissioned specifically for the programme, but fails equally to explore the relations between dance and the television medium in any more depth.

The piece, titled *Contrasts*, takes the form of a four-minute ballet and is set in front of a large sculpture which, on occasions, some of the dancers mount. The work is choreographed by Alfred Rodrigues and uses a small *corps de ballet* of six dancers and four principal dancers from the Sadler's Wells Royal Ballet. Yet regardless of the small numbers, with only one exception the ballet is filmed from a static, long shot that is face-on to the dancers. The movement content is abstract in style and draws upon established steps from the ballet vocabulary. There are only five or six cuts during the whole film and the camera remains static, so that there is little sense of dynamic or rhythm constructed through the camera and the editing. On one occasion a cut occurs between identical shots, which seems more like a 'jump' or fault with the film than an actual cut. As I mentioned earlier, the editing convention that is used in current television-making is that a cut marks a change of camera angle or shot size. The only change in shot size occurs with the final image, a close-up, in which the camera zooms in on one of the principal dancers to capture her smiling face.[20] It is clear that, irrespective of its televisual context, the piece is both choreographed and filmed with a stage perspective in mind. The dancing body is constructed as a 'complete body' and encoded within conventional ballet vocabulary and the hierarchical framework of the *corps de ballet* and principals.

In many ways *Houseparty*, which was filmed eight years later and billed as a 'ballet for television', provides a fascinating comparison to *Contrasts*. It was choreographed by Peter Darrell, produced by Margaret Dale and performed by dancers from the Western Theatre Ballet. Set to the music of *Les Biches* (1924), composed by Francis Poulenc, the work draws on similar subject matter to the ballet of the same name in the form of a hostess who entertains a variety of guests: these include her husband, her daughter, her son and his girlfriend, and the 'latecomer' who is a mysterious young woman in a glamorous cocktail dress. The piece is set in a trendy 1960s house and follows the events of a party that include scenes of: an abundance of smoking, mingling and social climbing; a sit-down meal; the latecomer throwing herself from the staircase; a game of blind man's buff; the daughter receiving unwanted sexual attention; and, finally, the latecomer being found dead, which brings a rather dramatic end to the evening.

Houseparty draws upon a pedestrian vernacular and both the dancers' movement and the camera work are tightly choreographed in relation to each other. As the piece utilizes a clear-cut narrative and somewhat stereotypical characters, the spectator is able to follow the thread of the work closely. Yet in comparison to *Contrasts*, many of the images are unconventional. First, the work is not choreographed or filmed from 'stage front', since the action takes place in several different spaces that include the lounge, the kitchen, the bedroom, a nursery, and a balcony. Many of the shots are filmed in close-ups which, quite notably, are not always of the typical facial variety. For instance, during the meal, feet are captured from under the table and hands are shown lighting cigarettes; at another point a pair of legs is caught walking down the staircase and some hands are seen tying a shoe-lace. Those shots which do focus on the face tend to display expressions that are over-projected. The dancers were possibly unprepared for the intimacy of the television medium and thus continued to perform in the way to which they were accustomed on stage.

The camera work is diverse and mobile. There are a number of overhead shots which focus on the action happening down below, 'zoomins' that draw the spectator into specific events or images and a 'panning'[21] motion, which follows performers as they travel from one group of people to another. This creates a vivid sense of dynamism. There is also a strong use of foreground and background, which often includes unusual perspectives. In one instance, someone puts down an ashtray that looms large in the foreground while the guests are seen mingling in the background. In another instance, the host is seen in the far distance inviting a group of guests through the door, but in the foreground, to the left of the frame, the hostess's torso comes into shot as she 'eyes up' the newcomers. This juxtaposition between the dominant profile of the hostess's torso in the foreground and the group of 'complete bodies' in the distance creates a striking image.

Unlike *We are your Servants*, and even *Cell*, the film cuts from shot to shot very regularly. In many ways the rapid cutting matches the upbeat and, at times, dramatic quality of Poulenc's music. The brisk style of edit creates another layer of rhythm in relation to the dynamism of the camera and the movement of the actual bodies; the pace is alert and sharp. The considerable number of cuts may also be accounted for by the fact that *Houseparty* was designed for television. In stage adaptations there is a fear that action may get lost or distorted due to the edit, hence there is a tendency to be restrained with regard to the number of cuts. With dance that is conceived for the camera, the choreographer designs the movement specifically for the medium; she or he can therefore create phrases of movement that are suited to a rapid series of cuts. Although, in comparison to *Contrasts*, the spatial and temporal possibilities of editing were well advanced in *Houseparty*, there are occasional instances in which a cut appears peculiar or inappropriate. For example, in one shot the hostess approaches the latecomer from the right side of the frame. In the next shot, however, the hostess stands inches away on the left side of the latecomer. Her sudden shift from one side to the other appears peculiar in editing terms, in that the the split-second cut does not account for the radical shift in position of the hostess in relation to the latecomer. This would suggest that, at the time of *Houseparty*, the conventions of editing were still in the process of being explored and developed.

Although *Houseparty* is a far cry from the unedited long shots of *Contrasts*, the spectator remains very centred. The more unconventional filming is cushioned by shots that clearly locate where the action takes place and what is happening. Due to the use of pedestrian movement, if a complete body is not depicted the viewer is able to predict what the remainder of the body is doing; there is no danger of missing out on any intricate footwork or hand gestures. This particular construction of the dancing body is somewhat paradoxical. On one level, the spectator is seeing a highly manipulated screen body; yet it is one that is deeply encoded with balletic stage conventions. For instance, the facial close-ups tend to linger across over-projected expressions and the pedestrian movement is performed with a balletic sensibility. Likewise, at the beginning of shots, the dancers are often posed in stillness and then suddenly move into action. They are obviously being cued, but the effect is very staged. There are other instances when principal characters are active in the foreground while characters in the background are statically posed; this device is similar to the tableau imagery of ballet.

In addition to standard choreographic devices, the piece is encoded with balletic stage conventions in terms of content: there are the intertextual references to *Les Biches* in the subject matter and the music; and when the latecomer throws herself from the staircase she is distraught and dishevelled, which is reminiscent of the 'mad scene' in *Giselle* (1841). It is also significant that, although *Houseparty* generally employs everyday movement behaviour and quotidian subject matter, which is notably in keeping with the type of realist images that are seen on television, the death of the latecomer at the end is a dramatic conclusion to the evening and is perhaps more in keeping with the tragic theatricality of many stage ballets. On the one hand, *Houseparty* utilizes facets of the television apparatus: the opening up of space due to multiple location filming; a fragmented body through the use of the close-up; and a clear sense of dynamism by way of camera movement and the rhythm of the edit. Yet *Houseparty* is also deeply inscribed by the codes and conventions of the theatre ballet in terms of subject matter, its narrative resolution, movement sensibility and performance style. Within this early example of dance for the camera, the body is simultaneously marked with screen and stage conventions.

It is apparent that different genres of screen dance deal with the filmic and televisual construction of the dancing body in distinctive ways. Although the examples under investigation clearly draw upon the codes and conventions of the television medium, the above case studies would suggest that each work constructs a dancing body that is largely based on already existing and well-established dance forms. *Flashdance* constructs a body deeply inscribed with popular conceptions of dance and, consequently, artistic innovation gives way to stereotypical imagery and mass consumer appeal. The Hellmann's mayonnaise and Cadbury's Twirl television advertisements similarly draw on popular representations of dance in order to promote specific social meanings in relation to the product. Although the music videos *True Faith* and *She Drives Me Crazy* employ the experimental work of Philippe Decouflé, the zany dance images are used to call attention to the musical structures of the songs and, consequently, to advertise the singles and the bands respectively. *Cell* meanwhile provides a representation of a dancing body that has spatial, temporal and dynamic consistencies with an already existing stage body. *Houseparty* is an attempt to create a work designed with the televisual apparatus in mind; yet the result is a work that constantly alludes to balletic conventions and the stage body. It would therefore appear that these examples participate in the fabrication of conventional dancing bodies, modelled on popular images or established vocabularies. They do, however, provide a useful theoretical foundation with which to analyse dance created specifically for the camera. Whereas Chapters 1 and 2 address the wider screen dance context, the remaining chapters concentrate solely on dance that is conceived for the camera.

3
Video Dance: Televisualizing the Dancing Body

An introduction to video dance

It is suggested in Chapter 1 that, although there has been a multitude of stage dance performances translated to the screen, it is apparent that there are difficulties with regard to this process. The live experience is a completely different phenomenon to dance that is located within a screen context. There is also a history of artistic tension between choreographers and directors in relation to adapting dance for television. Previously, this resulted from a lack of familiarity with the codes and conventions of each other's medium (Rubidge, 1993). Yet, even now, when choreographers and directors are experienced in translating dance from stage to screen, a sense of compromise remains: choreographers fear that the integrity of the movement will be distorted by the television medium, while directors are limited to shooting the dance in such a way that the choreography is fully visible. This is the case at least with 'faithful adaptations' of stage dance. In response to these practical and aesthetic difficulties, artists began to explore the possibilities of creating dance directly for the screen. These initial developments began in the mid 1980s in the United Kingdom. Although much of this nascent work was sporadic and underdeveloped, after a decade and a half there is a vast collection of dance pieces created for the camera. This body of work has come to occupy an experimental artistic territory in which innovative filming devices and cutting-edge imagery have come to the fore.

This new form of screen dance has been designated a selection of rubrics that include 'screen choreography' (Jordan, 1992), 'dance video creation' (Chaurand, 1993), 'camera choreography' (Brooks, 1993), 'dance for the camera' and 'video dance' (Maletic, 1987–88; Rosiny,

1994; Pritchard, 1995–96). The term 'video dance' is one of the most commonly used of these expressions and is employed throughout the remaining chapters.[1] What all these terms stress is a relationship between the medium of dance and a facet of the film or television media. Although 'video dance' is sometimes confusingly used to refer to any type of dance on video (de Marigny, 1991), including documentaries and stage adaptations, within the context of this book the term is employed solely in reference to this recently developed genre of screen dance that explores the 'dance–video' relationship.

In this book there is an intentional degree of flexibility within the term 'video dance'. The concept initially arose through a number of 'screen dance festivals'[2] as a means to delineate a category of work outside documentary films and stage adaptations. Since then, a new category has emerged that encompasses 'reworkings' of stage dance so that, to some extent, the original version is reconceived to take the film and televisual media into account (Meisner, 1991). It is apparent, however, that these definitions are vague and problematic. In some instances, 'reworkings' of stage choreography closely follow the codes and conventions of the original version, but may just happen to be filmed in a different location. At other times, a reworking of a stage piece may employ and manipulate the film or televisual apparatus to such an extent that the new version becomes a work in its own right and could only exist within a television context; consequently, it is far closer to the character of video dance than to stage dance.[3] It is not the aim of this book to formulate a rigid definition of video dance but to allow for a certain blurring of boundaries across these nebulous classifications. Therefore the concept of video dance can be taken to mean dance that is either originally conceived, or radically reconceived, for the television screen. For the purpose of this book, the primary concern is with work that explores the creative interface between dance and television in order to create movement images that could not be achieved in a stage context. Thus the emphasis is placed on innovation and experimentation (de Marigny, 1988; Meisner, 1991).

Although television is a visual form it is predominantly employed to pass on verbal information (Morley, 1995) and the medium is rarely explored as an aesthetic site in itself (Rubidge, 1984). Yet video dance has come to occupy an experimental artistic territory and it brings to light a plethora of pertinent issues and questions in relation to the theory and practice of dance and television. To commence, it is significant that although its output is fractional and funding is extremely limited (Meisner, 1991), this pioneering work has not been restricted to the fringes of art galleries and independent film festivals but is transmitted on public access, terrestrial television. It is paradoxical that video dance emerges from the margins of the 'art world', but it can potentially reach a vast public audience. Its increasing popularity is reflected in the number of screen dance festivals that have emerged (Schmidt, 1991). Yet the experimental nature of video dance has confounded dance critics and provoked strong reactions within critical circles. As identified in Chapter 1, the spectrum of responses ranges from those critics who condemn the technological mediation of dance through to those who have had to rethink completely the way in which they interpret and evaluate dance on television. In spite of these critical hiccups, however, video dance has attracted eminent practitioners from the fields of dance and television who champion its creative potential. This diversity of attitudes raises the question of why video dance has inspired diametrically opposed perspectives.

In order to investigate the impact of video dance on dance and televisual practices, Chapters 3 and 4 examine a range of video dance works. To allow for a broader spectrum of analysis, a cross-section of video dance pieces are considered rather than the work of a single choreographer or director. The research is based on two television series, screened between 1992 and 1996, that showed a range of video dance works: *Dance for the Camera* (series 2 and 3) (1995 and 1996) and *Tights, Camera, Action!* (series 1 and 2) (1993 and 1994).[4] As both series were screened on British television, I examine video dance within the framework of a domestic television viewing context rather than in terms of a screen dance festival.[5] The two *Dance for the Camera* series were screened on BBC2 and show a selection of works created by British-based artists. The series was produced by BBC television in collaboration with The Arts Council of England. All the works were commissioned by the producers and were originally conceived and choreographed for television. *Dance for the Camera 2* is made up of four, 15-minute works and *Dance for the Camera 3* consists of 12, five-minute pieces. *Tights, Camera, Action!* was made by MJW productions, an independent television company, and was screened on Channel Four. Each episode is a half-hour programme that shows between three and four works per slot:[6] *Tights, Camera, Action! 1* is made up of three episodes and *Tights, Camera, Action! 2* consists of four. *Tights, Camera, Action!* is slightly different from *Dance for the Camera* in that many of the works were created by international artists, and although some of the works were specifically commissioned for the programme, others were purchased from separate production companies.

The aim of Chapters 3 and 4 is to focus exclusively on video dance in order to begin to extract some of the predominant features that characterize this form. The basis of video dance is a creative exploration of the relationship between dance and television. Consequently it is appropriate that in any theoretical examination of video dance, this dual framework is acknowledged. For this reason an interdisciplinary approach is employed: Chapter 3 investigates how the televisual apparatus can intercept dance and construct a particular type, or types, of dancing body in video dance; Chapter 4 then addresses the implications of a postmodern dance aesthetic, out of which video dance has emerged, within the television context. This dual framework ensures that the codes and conventions that characterize each site and the ways that they can potentially act upon each other are accounted for. It is through this interdisciplinary methodology that fundamental questions about the character of video dance can be raised. The aim of Chapter 3 is to elucidate how the televisual apparatus acts upon dance within the video dance genre. The way in which televisual devices can manipulate and distort movement, influence and determine choreographic choices and enhance the dynamic quality of dance carries significant implications in terms of choreographic practice and the spectatorship experience. To commence, the following section investigates how the televisual apparatus participates in the construction of a 'video dance body'.

Manipulating the dancing body on screen

Two of the fundamental differences between dance and television occur in relation to notions of space and time. Whereas more conventional modes of screen dance attempt to construct a body that is commensurate with the spatial and temporal characteristics of the live body, it appears that video dance employs these phenomena in alternative ways. To begin with the concept of space, video dance explores certain camera perspectives to create spatial possibilities that could not be achieved on stage. The basic framing device of a camera enables a spectator to see compositional perspectives of an image that are only applicable to the film or television context. *KOK* (choreographer/director: Régine Chopinot, 1993) employs imagery from boxing and, at one moment, a performer's body is tightly framed in a mid-shot. The body then appears to catapult backwards, as if it had been punched, into a wider shot, and then rebounds off the ropes of the boxing ring to return to the medium-shot perspective. Due to the static, tight frame of the camera, the spectator has a clear sense of the diminishing and expanding body as it is thrown backwards and then lurches forward again. From a stage perspective the 'site'/'sight' of the body is imprecise and uncertain. Depending on where each viewer sits, the dancing body may appear to be in a slightly different location and its movement seen from a slightly different perspective. It is for this reason that, on film and television, unconventional perspectives work so well. The body can be framed in distinctive ways, yet the composition of the visual image always appears identical to each viewer.

As the television camera is able to frame shots in specific ways, in video dance this is exploited to the full. Through the use of close-up, the spectator has the opportunity to view dancing bodies from new perspectives and to focus on details of the human body that are barely perceptible on stage. Subtle muscular changes and facial expressions can become a part of the dance and, consequently, the dancing body shifts from a general body to a specific one. The close-range framing of particular body parts allows the viewer not only to see movement in precise detail but also to see images of disembodied or magnified body parts that are unusual to the eye. For example, at one moment in *KOK* there is a close-up of four heads tightly packed into the frame that dodge, pant and blow in unison as if ducking a series of punches. This twitching collection of heads is a striking and unconventional image.

The use of close-range filming is a popular practice in video dance and some pieces are filmed solely in close-up. *Hands* (choreographer: Jonathan Burrows, director: Adam Roberts, 1996) consists of a single, black-and-white close-up of the performer's hands situated on his lap, which execute a complex gestural motif. Similarly, *Monologue* (choreographer: Anna Teresa de Keersmaeker, director: Walter Verdin, 1994) is filmed in a single, black-and-white close-up, but in this instance the camera focuses on the performer's face, which shouts angrily into the camera. The choreography is located in the facial expression and the intensity of the close-up allows the spectator to see the furrowed brow, the elasticity of the mouth, and the pained, facial contortions that would be almost imperceivable on stage. Although *Touched* (choreographer: Wendy Houstoun, director: David Hinton, 1995), which is based on a drunken night out in a bar, is made up of many different shots, they are all from a close range, with the camera focusing on various people as they kiss, whisper, giggle, and nudge past each other in an intoxicated daze. What the close-up is able to do so well is direct the viewer's gaze to a particular area of interest. Unlike the prevalence of 'complete bodies' that characterize the faithful translation of stage dance to screen, the use of close-up presents a 'fragmented body'. Although close-ups are used to some extent in television adaptations of stage performances, these shots operate as a brief glimpse of detail, but ultimately return to a 'complete picture'. This type of visual coherence is far less common in video dance in which disembodied close-ups often have no apparent connection to the preceding and following shots.

In contrast to the close-up, the long shot is another televisual device that has been explored in video dance in order to create striking visual images. Although the long shot is sometimes considered problematic when filming dance, in that the body appears minute and distant, in several video dance pieces this has been effectively used to magnify the vulnerability or inconsequence of the body in relation to the vast landscapes that it occupies. For instance, in *boy* (choreographer: Rosemary Lee, director: Peter Anderson, 1996) a young lad races along a coastline, lost in his imaginary adventures. The long shot captures his elfin stature in relation to the great expanse of white sand. Similarly, *Keshioko* (choreographer/director: Saburo Teshigawara, 1994), a grainy black-and-white film, contrasts the diminutive figures of the performers against the enormous chimneys and lengths of cable in a deserted dockyard in the pouring rain. Again, the shot size portrays the miniscule bodies in relation to the dull industrial landscape. In these two instances it is not only the shot size that could not be achieved on stage but also the geographical location. The use of unusual environments and locations that could not be reproduced in a theatre setting constitutes another spatial feature that has been employed within many video dance works. For instance, *Dwell Time* (choreographer: Siobhan Davies, director: David Buckland, 1996) cuts between the carriage of a train and the roof of a building, *Le P'tit Bal Perdu* (choreographer/director: Philippe Decouflé, 1994) is set in a lush green field, and *Alistair Fish* (choreographer: Aletta Collins, director: Tom Cairns, 1995) moves from a swimming pool to a train. The possibility of situating dance in locations outside the studio/theatre setting is clearly an avenue for experimentation and the implications that such environments can bring to movement is explored later in the chapter.

In addition to location and shot size, the position of the camera can construct spectatorship positions that could not easily be achieved outside the film or television context. One instance of this, which has become a regular feature of video dance, is the 'top shot', in which the camera is placed directly overhead. This televisual perspective can distort and manipulate movement to the extent that it no longer resembles the same movement when seen from an upright position. The movement is even further mutated when it is depicted on the upright television screen. For instance, several top shots are employed in *Perfect Moment* (choreographer: Lea Anderson, director: Margaret Williams, 1993) as the dancers roll and wriggle across the floor. When these images are transferred onto the upright television set, the dancers appear to be suspended in mid-air and writhe across the screen in a malleable design. The lack of perceived three-dimensionality is somewhat jarring to the spectator's eye; yet its unexpectedness is also a point of fascination. A similar effect occurs in *KOK* when the performers sprint back and forth across a boxing ring. As the action is filmed from a top shot, the bodies are distorted into a mutable pattern of constant flux. In *Mothers and Daughters* (choreographer: Victoria Marks, director: Margaret Williams, 1994) there are several top shots. At one moment the performers drag, coax and push each other across the screen and the top shot allows the spectator to see this combination of forceful and recalcitrant bodies from a new dimension: limbs shoot out, shoulders jostle and heads bob back and forth. At another point, the mothers lie in a diagonal pattern across the floor while the daughters stand above them. On the upright television set, the mothers appear to be suspended across the screen and, in contrast, the daughters take up a minimal degree of space from this overhead perspective. This play on spatial logic gives the bodies an almost abstract quality. It would appear that video dance completely abandons the fixed, upright spectatorship position that characterizes theatre dance and instead focuses on constructing innovative viewing experiences that expose perspectives of the dancing body which are striking and unconventional.

The way in which a shot is framed can also challenge the spatial logic of dance. In *Codex* (choreographer/director: Philippe Decouflé, 1993) the opening image is turned upside down. It consists of three performers, dressed in black rubber wet suits, diving flippers and small black hats, in front of a plain, pale-blue background. The central performer balances on her or his head and faces away from the camera, while the two performers on either side stand upright and face the camera. Due to the plain backdrop and the upturned image, the performers seem to be literally suspended in the middle of the screen like a chain of paper cut-outs. As the performer in the central headstand position slowly lowers her or his feet to the ground, the performers on either side simultaneously bend down towards the floor in a 'flat back' position. With the two-dimensional screen, the performers almost appear to fold in half. The central performer eventually comes to standing position while the other two performers slowly move into a headstand. The result is that their positions have been reversed and the whole image appears to have turned upside-down. The two-dimensionality of the television screen and the 180-degree rotation of the image clearly distorts the spatial logic of the movement.

A similar sense of spatial manipulation can be seen in *Keshioko*. There are several shots in which either the camera is placed on its side or the video image is turned sideways so that the performers appear to dance on a vertical plane. Although the idea is simple, with the 90-degree rotation the movement appears strange and the logic of gravity becomes distorted. In much the same way that video dance disregards the fixed spectatorship position that characterizes stage dance, it also abandons the spatial limitations and conventions of the live body: the body becomes a site of experimentation for the televisual apparatus.

Some genres of screen dance, as can be seen in *Cell*, attempt to counteract the two-dimensional image through constructing an illusion of three-dimensionality; yet several video dance pieces actually emphasize the two-dimensionality of the image to create alternative perspectives. The two-dimensionality of the television image is explored in *Codex* in which images that could not be achieved on stage come into existence. Through the use of bold geometric designs, plain, contrasting colours, and a specific use of background and foreground in relation to the camera, *Codex* cleverly manipulates the scale of the body. Perspective can become completely distorted due to the two-dimensional screen. During a scene, although one body is in the foreground and the other is in the background, the two-dimensional screen appears to situate an enormous body beside a minuscule one.

This type of spatial disorientation, which results from the position of the camera in relation to the body, is also explored in other video dance works. In a stage setting, due to the static, upright viewing position of the spectator, she or he has a clear sense of whether the performer's body is vertical or horizontal. In video dance, as the camera is able to take up any position in relation to the performer, this can sometimes disrupt the spectator's sense of spatial logic. At one moment in *Cover-up* (choreographer: Victoria Marks, director: Margaret Williams, 1996), the performer kneels down and wipes away the 'snow-covered' ground to reveal a black surface on which she marks out the shape of a body. The camera cuts to a top shot, so that the image consists of the performer, huddled over a black 'body', surrounded by snow. With the camera still in an overhead position, she then lies down on top of the body shape. The following shot reveals the same image, but within seconds the spectator realizes that the spatial logic has been completely transformed. It becomes apparent that the camera is now face on to the image and the performer is upright, as she suddenly turns and runs straight through the black body, which is now in fact a body-shaped fabric, with a slit down the centre, on a white backdrop. The change of camera location and alteration of the performer's position completely undermines the spectator's spatial orientation.

A similar example is in *Waiting* (choreographer/director: Lea Anderson, 1994), in which three performers, dressed identically in long, red wigs and grey dresses, sit on black chairs which are placed against a white studio setting. As the performers appear to be sitting upright on the television screen, the spectator does not even stop to question whether the performers and the camera are upright, although this is not actually the case. The chairs are, in fact, built into a frame and are in a parallel position to the ground, while the camera is placed either overhead, so the dancers appear to be sitting upright, or on its side, so that they are in profile. The result is that as the dancers begin to traverse over and around the chairs, their movement appears peculiar and distorted as in actuality they are working against the pull of gravity. Their bright red hair flies out at 90 degrees to their heads and their bodies appear to be weightless. The image is disorienting to the viewer.

In much the same way that video dance explores and manipulates the spatial possibilities of television, it also investigates and exploits the temporal characteristics. Unlike more conventional modes of screen dance, video dance does not necessarily set out to fabricate an illusion of temporal linearity. In many instances it draws attention to and plays on the 'constructedness' of televisual time. The use of fast motion is one example of this. In *Codex* the image is speeded up at one point to such an extent that the dancing figures move at a physically impossible rate. Through this acceleration technique the spectator is able to see a new level of kinesis. This type of 'fast-motion body' in dance also carries certain choreographic implications in that some types of movement are more effective at a rapid speed than others. For example, small and subtle movement is likely to vanish at such an accelerated pace whereas full body movement has a greater element of clarity when speeded up.

In *Codex* this fast motion device is cleverly used to confuse the spectator's sense of temporality. At one point two of the performers stand in the centre of the space and appear to move at normal speed as one of them folds over into a headstand and the other supports this pose. Within the same image, however, another group of performers run around the duo at an impossibly rapid pace. The effect is both bizarre and fascinating, yet it is based on a simple idea. During filming, the two central performers moved extremely slowly while the remaining performers moved at normal running speed. The image was then speeded up, either during filming or post-production, making the central performers seem to move at normal speed and the other performers flash past the screen at an impossible rate.

At the opposite end of the temporal scale, slow-motion filming devices are used to construct movement that could not be achieved outside the film or televisual apparatus. Several pieces of video dance have sections of action played in slow motion, allowing the spectator to see facets of movement that are barely perceptible when played in real time. For instance, in *Monologue* this lengthening of time facilitates a detailed examination of the performer's face as she shouts aggressively at the camera: the crinkles in her face, the curvature of the mouth and the way in which she blinks are seen in painstaking detail. Not only can slow motion allow for a more precise viewing of the body but it also constructs a different quality of movement. *Relatives* (choreographer: Ishmael Houston-Jones, director: Julie Dash, 1994) is a solo improvisation performed by Ishmael Houston-Jones in response to the quality of his mother's voice. During one particular sequence, set in his mother's back garden, the image alternates between normal speed and slow motion. The everyday speed gives the movement a pedestrian or utilitarian quality, whereas the slow motion creates a sense of suspended motion, a defiance of gravity and a dream-like ethereality.

Unlike stage dance, screen dance is not bound to temporal linearity, and video dance exploits this in a variety of ways. In *Le Spectre de la Rose* (choreographer Lea Anderson, director: Margaret Williams, 1994) a whole section is played in reverse, so that at one moment a dancer appears to be miraculously sticking petals onto a rose rather than plucking them off. Although the movement is relatively simple, the use of reverse motion subtly distorts the physical action. Similarly, in *Codex*, at the end of the first section, a performer executes a series of 'backward handsprings' across the television screen. As the image is played backwards the performer appears to execute odd-looking 'forward handsprings' and moves in a manner that is physically impossible outside the film or television medium: the execution of those particular take-offs and landings in reverse are simply not possible for the live body to mimic.

There are also other ways in which temporal linearity is subverted in video dance. In *Cover-up*, a still frame device is used so that during a solo, at intermittent points, the performer's movement is momentarily frozen. What makes this sequence particularly interesting is that this suspension occurs during certain movements, such as aerial or off-balance steps, which could not be achieved outside the film or television apparatus. The screen body is able to execute the seemingly impossible. In *Man Act* (choreographers: Man Act, director: Mike Stubbs, 1996) a 'looping device' is employed to repeat identical sections of material. The work examines father–son relationships and one particular sequence, in which the camera appears to circle around two men who sit opposite each other, is repeated over and over again. This type of precise replication can never be achieved on the live body. Whereas other forms of screen dance attempt to construct the sense of temporal linearity that characterizes live performance, video dance experiments with and disrupts perceived notions of time.

In addition to the manipulation of spatial and temporal possibilities in television, various other televisual effects are employed in video dance to create dancing bodies that could not be realized on stage. One example is through the use of colour. *Elegy* (choreographer: Douglas Wright, director: Chris Graves, 1994), *Man Act* and *Touched*, to name but a few, are all filmed in black and white. Through the use of televisual technology, the body can be re-presented in different shades and contrasts of colour in order to create bold, striking images. The visual quality of the image can also be affected in various ways. For instance, *Kissy Suzuki Suck* (choreographer/director: Alison Murray, 1993) employs a rough, grainy image so that the texture of the performers' bodies appears coarse and unrefined. Not only is this in contrast to the smooth, glossy bodies that are generally seen on television (Foster, 1992a), but also the 'crude' quality of the image stylistically reflects the 'sleazy' theme of prostitution on which *Kissy Suzuki Suck* is based. A similar device is through the use of focus. In *Attitude* (choreographers: RJC Dance Theatre, director: Anne Parouty, 1996), fleeting sections of the piece are filmed out of focus; the body is no longer a high-definition image but an indistinct blur, impossible to pin down. Through this technique television has the capacity to distort the boundaries of the body: no longer a solid mass, but nebulous and intangible. This instability of the body can also be seen through the use of a 'dissolve'.[7] In *boy*, a close-up of the child's brisk, wiggling fingers dissolves into a medium shot of him prowling through the sand dunes. Within television, the body does not constitute a permanent or stable entity, but is transient and unpredictable.

The above analysis suggests that the televisual apparatus is able to construct dancing bodies that could not be replicated on stage: fragmented bodies, magnified bodies and minuscule bodies; bodies seen from unconventional perspectives; unpredictable bodies that undermine spatial and temporal expectations; and bodies moving in ways that are physically impossible outside the film and television context. Although some of these characteristics, such as the use of close-up to focus on specific areas of the body, are employed in screen adaptations of dance, in general the intention of these works is to minimize or underplay the presence of the televisual apparatus through such devices as 'invisible editing', slow and sustained camera work, and conventional filming techniques. Conversely, as video dance sets out to explore the creative potential of dance and television, innovative filming and unconventional images are predominant features of the genre. Indeed, video dance actually highlights and plays upon the workings of the television medium, and this manipulation and experimentation carries a number of implications for the dancing body. It could be suggested that, through the televisual apparatus, the limitations of the material body are radically challenged. Perceptions of what a dancing body can do are extended as the spectator sees bodies defy gravity, travel in slow motion and reverse certain movements, which in reality would be physically impossible. In many ways this would suggest that video dance allows for a more versatile dance body.

There are also other implications for the dancing body that apply not only to video dance, but to all screen bodies. With dance on film and television, any drastic mistakes can be edited out or re-shot. The potential for a 'flawless dance body' has been seen as a criticism, however. Barnes (1985) insists that the sense of danger or failure that comes from watching live performance is never present in screen dance as the 'perfect take' is always used. On one level this is true. The fact that screen dance can potentially allow for the perfect performance means that there is little tension in respect to whether the dancer will successfully execute her or his steps. On the other hand, an alternative mode of suspense can be constructed through film and televisual techniques. For instance, through employing certain camera positions, a fall can appear to be far more intense than it actually is, or elements of a chase can be dramatically heightened through the use of rapid editing. It has also been suggested that the dancer's body tires of doing repeated takes for television and, as a result, the spontaneity of the single performance is lost (Eisele, 1990). In response it could be argued that the dancer's body also has 'off' days in the theatre and spontaneity may equally be lost if a particular performance has run for several months.

To return solely to video dance, part of the fascination of live dance performance is the sight of virtuosic bodies achieving movement feats that are simply not possible for the untrained, or even the average, dancing body to do. This notion is perhaps reflected in *Flashdance* in which the dancing body is inscribed with virtuosity, risk and technical brilliance. Although it is a highly mediated screen body, it is constructed in a way that conforms to the popular conceptions of the live dancing body as slim, youthful and athletic. Another criticism of video dance, then, might be that the televisual technology has somehow negated this pleasure of witnessing the virtuosic body since the body in video dance is able to do all manner of things which the material body, and even the virtuosic dancing body, are unable to do. In defence of video dance, it could equally be argued that the form offers a more democratic dancing body. For instance, one does not need to be a certain age or body type in order to become a 'spectacular' body on screen. The possibility of moving in fast motion, or defying gravity, is accessible to any body. Of course there are stage performances in which non-virtuosic dance bodies are also used. What makes video dance so distinctive is that the aesthetic value of virtuosity is negated as the technology of television can enhance and extend the possible movement ranges of the body to such an extent that, on screen, it has no physical limitations. In many ways, innovation overrides virtuosity.

The bodies constructed within video dance are not simply screen bodies but specifically stylised bodies that are explored and manipulated by the televisual apparatus to create innovative and striking images that can exist only within this context. In order to differentiate this body from the other dancing bodies that exist on screen, the notion of a 'video dance body' can usefully recognize this distinction. This is not to suggest that the video dance body is a single, fixed body, but a generic category that embraces a multiplicity of televisual possibilities evident within video dance. Now that some understanding of the relationship between dance and the televisual apparatus in video dance has been established, the following section investigates the type of movement that has come to the fore.

Choreographic content in video dance

In relation to choreographic content, a number of discernible trends and characteristics have begun to emerge through the video dance form. Therefore the focus of this section is directed at the ways in which movement in video dance may be influenced or determined by

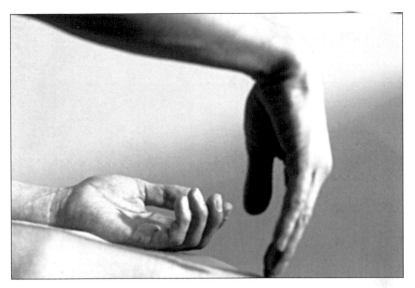

1 *Hands* Choreographer: Jonathan Burrows, Director: Adam Roberts, 1996 ©
Arts Council of England and BBC Television.

2 *Alistair Fish* Choreographer: Aletta Collins, Director: Tom Cairns, 1995 © Arts
Council of England and BBC Television.

3 *Touched* Choreographer: Wendy Houstoun, Director: David Hinton, 1995 © Arts Council of England and BBC Television.

4 *Drip* Choreographer: Matthew Bourne, Director: Frances Dickenson, 1996 ©
Arts Council of England and BBC Television.

5 *Dwell Time* Choreographer: Siobhan Davies, Director: David Buckland, 1996 © Arts Council of England and BBC Television.

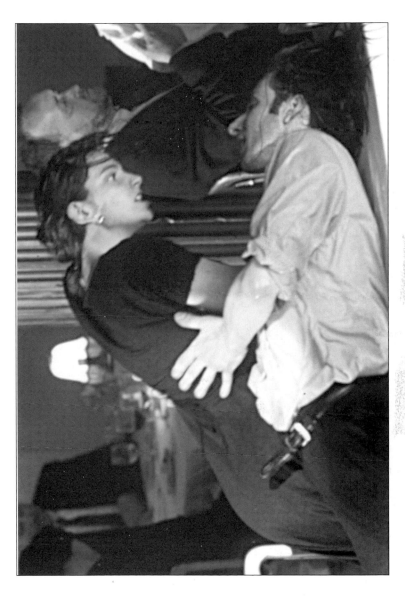

6 *Storm* Choreographer: Aletta Collins, Director: Tom Cairns, 1996 © Arts Council of England and BBC Television.

7 *Man Act* Choreographers: Man Act; Director: Mike Stubbs, 1996 © Arts Council of England and BBC Television.

distinctive feature of *Mothers and Daughters*, which situate the performers' bodies within unconventional perspectives and abstract designs;
the effects are riveting. In *Waiting*, the chairs are flipped 90 degrees
onto their backs and the camera placed directly above them so that, as
the performers begin to move around, the sense of gravity becomes distorted. The action is simple, yet the movement quality appears weightless and bizarre. *Kissy Suzuki Suck* is meanwhile filmed in a grainy,
black-and-white, hand-held camera style. The jumpy camera work and
rough texture of the film give the everyday action a visual and kinetic
quality which it would otherwise not have.

It is perhaps not surprising that such a degree of pedestrian movement is used in video dance. In comparison to the filmic image, the
television screen depicts a relatively poor quality overall image, and
thus often relies on its audio component (Monaco, 1981). In some
ways video dance has to compensate for this through such devices as
bold colours, clear composition, ample close-ups and movement that is
not too elaborate. This may be one of the reasons that the simplicity of
pedestrian movement works well on television. It may also be relevant
that the television viewer is used to seeing all kinds of pedestrian
action on television, ranging from teadrinking to racing down the
street. The use of pedestrian activity is easily identifiable in video
dance and perhaps provides a more accessible route for the 'nondance
expert' to engage with a work. Since the television camera is suited to
picking up on detailed, close-range activity, the slightest of actions
becomes significant. Subtle, commonplace action can become imbued
with a wealth of visual and kinetic information once recontextualized
within the television medium. As Lockyer notes, 'There is a greater
immediacy with television, raising an eyebrow or turning the head
takes on the importance of a grand jeté' (in Craine, 1995, p. 6). The
simplicity of pedestrian action facilitates a certain degree of manipulation in terms of framing, camera movement and editing. If movement
is too elaborate it becomes more difficult to find innovative ways of
filming for fear that its intricacy may get lost.

Some of these considerations have clearly influenced other types of
movement employed in video dance. For instance, the camera's ability
to focus on movement through the close-up has resulted in the use of
small, detailed activity in video dance that would be barely perceptible
in a stage context. In *Mothers and Daughters*, for example, a recurring
motif is made up of separate close-ups in which a mother and daughter
respectively raise a single eyebrow in an identical fashion. *Le P'tit Bal
Perdu* meanwhile employs a complex and intricate gestural vocabulary

based around the hands and face, which is complemented by a detailed use of close-up. This suggests a shift towards choreographing whole dances on specific body parts rather than the body as a whole, which perhaps offers new possibilities for injured dancers or nonfully abled dancers. This is clearly a move away from the virtuosic dancing body as promoted in *Flashdance* and televisual adaptations of stage performances.

The idea of choreographing complete dances on specific body areas has come into practice with several video dance pieces. As previously stated, the choreography for *Monologue* is located in the performer's face as she directs a pained, verbal attack towards the camera. Similarly, *Hands* features a single close-up of a pair of hands situated on the performer's lap. Through precise spatial designs and a set of simple gestures, which include clenching a fist, the middle finger marking out a line along the thigh and the thumb rotating across to the little finger, the performer constructs an elaborate dance located within the boundaries of a particular shot size and a specific area of the body.

In contrast, there are certain works that utilize the complete body and retain a more discernible dance element; works which have shaped the movement to suit the relatively poor quality image of the television screen. *Keshioko* is filmed in black and white, at a deserted dockyard in the pouring rain. Many of the images are filmed in long shot so that the dancers' bodies appear minute and distant against the dull, grey landscape. For the movement to be successfully seen, the choreographer has employed simple, linear actions that consist of straight arm extensions, 90-degree leg kicks to the side, a wide 'second position' and sideways lunges. The movement is predominantly two-dimensional, using height and width rather than depth, which complements the flat television screen. Similarly, the first section of *Codex* contains a large group of performers dressed from head to toe in black. Under normal circumstances, such a vast group could easily become very muddled on the small television screen. The movement content for this section, however, is made up of diagonal arm positions, sideways tilts, upper-back curves, '*demi-pliés*', running, tapping feet and shrugging shoulders. These clear and simple lines are easily identifiable on the two-dimensional screen.

Some video dance pieces construct movement that could only ever exist on screen. *Topic II/46 BIS* (choreographer: Sarah Denizot, director: Pascale Baes, 1993) employs a 'pixillation' technique'.[9] The pixillation device is able to cut out sections of time in a film so that the performers can appear to move from one position to another without any form of transition. For instance, a dancer may pose in one position and then move into another pose several inches away from the original one; however, as the pixillation device can omit the transitional movement, the performer appears to 'skate' along from one pose to another in a slightly jerky fashion. For instance, *Topic II/46 BIS* is set in the streets of Prague; the dancers twirl, sit and lie down, pose against walls, ascend a flight of steps and get dragged behind a trolley as they appear to slide continually across the ground, without any pauses or transitions. A style of movement is produced that could not be replicated outside the televisual context.

Another example of movement that can only be achieved on television can be seen in the formalist piece *Tango* (choreographer/director: Zbigniew Rybcznski, 1993), which employs an animation technique in order to create an accumulation of bodies that repeatedly enter and exit a room as they perform their day-to-day activities. The film is set to a piece of tango music and the camera remains face-on to the room throughout. The piece begins as a small boy outside throws a ball through the window. He climbs through the window, collects his ball and climbs back out of the window. This image is then repeated so that his action is identical to the previous time, but as this is happening a woman enters holding a baby. She sits for a moment at the chair breastfeeding, then puts the child in the cot and exits. This pattern continues so that more and more people are introduced, until the room is full of people entering, exiting and performing tasks: a jogger does a headstand on the table, a man carries a lavatory bowl across the room, a thief enters through the window, a couple copulate on the bed, and so on. Not only is the pedestrian action, identically repeated time and time again, compelling to watch, but so too the fact that the piece could only take place on two-dimensional bodies. This comical work clearly exemplifies the possibility of movement that could not be achieved outside the television context. In reality it would be impossible to fit so many moving, three-dimensional bodies into a single room, thus to achieve this multi-layered accumulation it is necessary to superimpose one image over another. As a result, *Tango* can only ever take place on the two-dimensional bodies of television.

It is possibly because television is potentially accessible to the diversity of a 'broadcast' television audience, rather than a specialist, theatre dance audience, that some video dance pieces have drawn on popular or social dance styles. For instance, *T-Dance* (choreography: Terry John Bates, director: John Davies, 1996) is based on the senior citizens who attend a ballroom dancing club, and *Horseplay* (choreographer/

director: Alison Murray, 1996) draws on the 'street' and club dance trends of youth culture. Indeed, Alison Murray, who choreographed and directed *Horseplay*, clearly aims to employ movement styles that may be accessible to the broadcast audience. She states:

> I'm really interested in all the references of film and television and working within that kind of context, which I think is one thing that might set my work apart from a lot of the other dance films, because they're coming purely from a choreographic, dance context, and they're aiming at a dance audience. Whereas I'm trying to aim my work more at a TV audience who don't necessarily participate in dance in the more sort of theatrical environment.[10]

The new or alternative environments in which some video dance pieces are situated facilitate movement that could not easily be reproduced in a stage context. For example, as *boy* is situated on a beach, much of the movement derives from the performer's physical relationship with the sand. In several instances he executes dare-devil leaps from the peak of a sand dune, then sinks his heels into the soft, fine particles, slithers across the dunes in a snake-like fashion and sits manipulating the damp sand with his fingers. Another instance is in *La La La Duo No 1* (choreographer: Edouard Lock, director: Bernard Ebert, 1994). The work is set in a dance studio and, towards the end of the piece, the room slowly fills with water. One of the performers floats on the surface of the water while the other is depicted underwater as she paddles her feet to swim out of shot. This buoyant quality of movement clearly results from the work's environment. Another example of this can be seen in the swimming pool shots of *Alistair Fish*.

The use of alternative locations for the presentation of dance and the exploration of pedestrian and gestural movement vocabulary, technologically mediated movement, and popular or social dance styles, undoubtedly have some impact on the type of performers employed within video dance. A significant feature of the video dance genre is its use of non-dancers or unconventional body types for dance. For instance, with the exception of one person, *Mothers and Daughters* utilizes a group of performers who had no previous dance training and who range from a small baby through to women in their fifties. *Man Act* meanwhile focuses on a group of males of a similar age range who also had no previous involvement with dance. It is perhaps because these performers had no formal dance training that quotidian movement was seen as a more appropriate vehicle with which to tackle the subject matter of these two pieces. This is a stark contrast to the narrow representation of the slim, youthful and athletic dancing body as promoted in popular Hollywood films such as *Flashdance*.

The use in video dance of certain body types rarely seen in stage dance is also reflected in *boy*, which focuses on a young lad, and in *Bruce McLean* (choreographer: Bruce McLean, director: Jane Thorburn, 1994), which is based on the actions of a portly, middle-aged man. As video dance is particularly concerned with a creative exploration between dance and television, this is perhaps why existing vocabularies and technical dance steps take a secondary position to movement that is constructed by, or in complement to, the televisual apparatus. In *boy* the movement of the camera and the dynamic of the edit create a strong sense of imaginary adventure in relation to the boy's playful movement; and in *Bruce McLean* the rhythm of the edit imposes a frenetic quality onto the performer's everyday action. Another example is in *T-Dance*. As the piece is based on the recreational setting of a tea dance, it employs a group of senior citizen dancers. Yet although young boys, middle-aged men and the elderly may be relatively unconventional performers within the stage dance context, they are well represented within television. Therefore, the wider variety of body images within video dance is perhaps a conscious effort to appeal to, and reflect, the diversity of people who constitute both the television audience and television texts.

There are, of course, several examples of video dance that solely employ traditional, theatre dance styles. For example, *La La La Duo No 1* is based around a highly physical contact duet, *3rd Movement* (choreographer: Matthew Hawkins, director: Deborah May, 1994) manipulates and plays with a classical vocabulary, and *Elegy* is rooted within a contemporary dance technique. It could be suggested, however, that one reason for the prevalence of gestural and pedestrian movement in video dance is possibly tied in with the links between high art and authenticity, and mass culture and reproduction. Whereas on stage the spectator is able to see the authentic, live performance, dance on television may be reproduced thousands of times over and is accessible to anyone. Physical virtuosity can meanwhile be constructed and doctored through the televisual apparatus by way of clever camera angles, carefully designed editing and special effects. This is perhaps why video dance has often moved away from highly trained dancers and technical 'art dance' vocabularies in favour of pedestrian and gestural activity, social and popular dance styles, and movement that exists by way of the camera and the edit. Notions of authenticity that characterize the live performance do not hold such priority in the world of the electronically produced image. Since video dance is motivated by artistic experimentation, it is hardly surprising that unconventional styles of movement are employed.

To some extent the attempts to develop innovative relationships between dance and television have led to a reconceptualization of what constitutes dance. For instance, film material of the movement of internal muscles and organs is used in *Joan* (choreographer: Lea Anderson, director: Margaret Williams, 1994). This brings a whole new meaning to the concept of dance when the possibility of 'involuntary corporeal movement' is taken into account. It is only though the use of endescopic cameras that this type of 'movement' may reach a viewing public and be situated within a choreographic context. A similar example is with *Monologue* and, indeed, many of the video dance pieces that employ everyday motion[11] or movement that is barely perceptible in a stage setting. In essence, *Monologue* is a piece of facial choreography. The movement quality derives from the earnest expression of the performer, her increasing frustration and the facial contortions that derive from her furious disposition. This is augmented by the slow-motion filming which allows the spectator to see facets of a 'shouting face' that would not be seen under normal circumstances, the single, tight close-up and the accompanying Monteverdi soundtrack. The relationship between these various components is what removes the movement in *Monologue* from the everyday and relocates it within an aesthetic context. It is perhaps because video dance has challenged existing boundaries of what 'dance' is that it floors so many dance critics in their appraisal of video dance.

It is apparent that movement that is effective on stage does not necessarily translate to television, and movement that is compelling on television may not be particularly striking or successful on stage. This obviously carries choreographic implications for video dance because of the importance that movement is conceived and designed in relation to the television context. Indeed, it appears that certain types of movement have already begun to characterize video dance: pedestrian and gestural movement, isolated and detailed action, simple, geometric motion, technologically enhanced movement, social and popular dance styles, and movement that complements the location are some of the emergent features of video dance. This consideration of the television context would strongly suggest that physical movement in video dance is inextricably linked to the televisual apparatus. The televisual mediation of dance has the potential to create innovative and striking forms of dance. The construction of movement in video dance, however, is not only dependent on the size of the screen, the quality of the television image and the possibilities that special effects, location, shot size and camera position can bring to dance, but also on the movement of the camera and the nature of the edit.

The dance of the camera and the cut

In addition to the physical movement of the body there are two features of television that can construct a sense of motion and augment the dynamic quality of the image, independently of the actual body: the camera movement and the cut. For instance, even if a body is static it can move around the screen depending on the movement of the camera. If the camera tracks backwards, a body would appear to recede into the background, and if a camera pans to the left, a body would appear to travel across to the right of the screen. Similarly, the style of the edit can significantly contribute to the dynamic sensibility in video dance. A rapid cut can construct an illusion of high-speed action, whereas an absence of cuts can create a sustained calm. As the role of the camera and the style of the edit are inextricably linked to the quality of motion in video dance, it could be argued that it is not just the physical body that constitutes the 'dance' but the triadic relationship between the moving body, the camera and the edit. Consequently, in video dance, the camera work and the style of the edit are essential components of the dance itself.

In *L'Envol de Lilith* one section centres on the caliph's harem. There is little physical action in this scene as the women lie around playing musical instruments, chatting and decorating each other's hair. The movement quality derived from this scenario comes more from the lingering, voyeuristic tracking shots that pass by the various women as they indulge in their luxurious lifestyle than from the actual movement of their bodies. Another example is in *Horseplay*, which focuses on the playful activities of three young women as they 'knock around' a quiet stretch of urban wasteland. Their physicality is upbeat and energetic as they push each other around, pull faces, emulate great footballing moments and try out various street dance styles. The camera work in *Horseplay* clearly matches their high-spirited antics. During some shots the camera is attached to a boom-pole so that it can swoop in and around the dancers. It cleverly captures their playful dynamic as it sets up a shot from overhead but then dives down to a low angle image, or else circles around the three women and then flies upwards until they are framed from a completely new perspective. At one moment the dancers run backwards and the camera jerkily reverses along with them; at another, the dancers leap back and forth across each other and the camera reflects these precarious moves as it pans along in a low-angle shot. The energetic camera work clearly contributes to this mischievous scenario.

As with camera work, the style of edit can play a major role in the type of dynamic that is constructed. For instance, in *KOK* a series of shots occurs during which the performers stand in a selection of boxing poses. Although the bodies are actually static, the rapid edit of these shots creates a sharp, staccato rhythm. Since the aim of video dance is to create innovative relationships between dance and the television medium, there has been a certain amount of experimentation with regard to the edit. At one end of the editing spectrum are works that have a cut every second or so, and at the other are completely unedited pieces of film. For example, in the *Bruce McLean* video a sense of twitching, nervous energy is constructed; this is less to do with the movement dynamic, which derives from basic pedestrian gestures, than with the rapid editing that persistently jumps back and forth across different shot sizes. A similar example can be seen in *Pace* (choreographer: Marisa Zanotti, director: Katrina McPherson, 1996). The dancer moves around a rehearsal studio in a relaxed, release-technique style; yet the high-speed edit gives the work a frenetic, high-energy quality. In stark contrast, the video dance pieces, *Hands* and *Monologue* both consist of a single, unedited shot. In the case of *Hands*, the observant, unedited close-up clearly complements the attention to detail and careful execution of the gestural sequence. In *Monologue*, however, the intensity of the single close-up only serves to magnify the wild, impassioned speech of the performer; the effect is almost claustrophobic.

In order to explore further the role of the camera and the edit in video dance, the remainder of this section makes a detailed analysis of two pieces that demonstrate this well. The first, *Touched*, is described as 'a choreography of close-ups' in its publicity notes, which takes into account the role of the televisual apparatus in the construction of movement. The film is shot in black and white and is situated in a bar full of people. The movement is based on the pedestrian action that takes place in this type of social context and, as the publicity statement suggests, the whole film is shot in close-up. Although the piece does not follow a narrative as such, various 'movement dialogues' develop and certain 'characters' emerge. The opening shot focuses on a pair of hands, which are then met by another pair of hands. The camera slowly moves up the body to reveal a face; yet various hands, which touch, jostle and stroke, still appear in the frame. Just as another face appears, the camera begins to move back down the body to return to the hands. The slow, subtle camera work marks out this dialogue between body parts while its actual route creates a strong spatial pattern. The camera then continues to pick up on other gestures, incidents and social interplay.

The rhythm of the piece is not only dependent on the speed and dynamic of the physical movement but also on the quality of camera movement and frequency of the cut. For example, at one moment the camera pans along with a pair of hands with wiggling fingers, as if about to tickle someone. The slow camera movement creates a gradual build of suspense. Just as the hands grab someone by the waist the shot suddenly cuts to a face that jumps in surprise. Another head, which presumably belongs to the tickling hands, meanwhile playfully snuggles into the shot. What makes this scenario work so well is the suspense of the slow, sustained tracking shot followed by the sudden cut which neatly captures the shock of the woman who receives the unexpected tickle.

There are other instances where the camera and the edit significantly contribute to the quality of the movement. During one scenario the camera is placed as if on one side of the bar while a drunken man is placed on the other. He totters back and forth along the bar and occasionally slumps towards the camera. The camera pans back and forth with him along the bar, which heightens the lolling quality of his movement. As he staggers to and fro the camera remains static so that he looms large in the foreground at one moment and then radically diminishes as he lurches backwards. These shifting dimensions capture the askew perspectives that come with a drunken night out. As part of this same scene the camera also picks up on a drunken woman whose head falls heavily back, and curves round to lead her body into a full circular turn, to then return to standing. Again, the camera work highlights this lopsided turn and follows her body in a similar circular motion. The woman slams her glass down on the bar and the camera cuts to a close-up of the glass. Throughout the whole drunken section, in which a number of other characters emerge, the edit points are frequent, creating a sharp and pacy rhythm. The lolling camera work and jumpy cutting cleverly recreate the sense of incoherent fragments and dizzying perspectives of an alcohol-fuelled evening.

Due to the regular cuts, the use of close-ups, the size of the movement and its pedestrian nature, much of the choreography in *Touched* consists of short, and almost understated, sequences. For instance, a shot may simply consist of a change in facial expression or a scratch of the head. Vignettes of material replace long sequences of movement. What gives the film a sense of whole is the way in which the short, individual shots are edited together. Another result of the extended use of close-ups is that it becomes difficult to locate which body part belongs to whom. At times the viewer only sees a series of hand or feet motifs, without any relation to the rest of the body. This suggests an element of instability with the body in video dance. Not only is it able to surpass the capabilities of the material body but its boundaries are not always clear. The viewer cannot necessarily see the body's beginning and end, and hence there is no concept of a coherent whole. It is perhaps this device that gives *Touched* a sense of enigma. It is up to the viewer to try and construct the fragmented close-ups into a complete scenario.

The second of the two films, *Le Spectre de la Rose*, is performed by five men dressed in rose petal suits, and takes place in the grounds of Nymans Gardens and in a white studio. The opening shot creates a sense of motion without a body even being present. A 'steadicam'[12] is employed and rapidly traverses a large expanse of grass. The bumpy movement of the camera and its focus on the grass creates a sensation of someone running. As an old ruined building comes into sight, the camera stops and pans from side to side, as if someone's eyes are scanning the building. The sense of motion is clearly derived from the technology. The ability of the camera to create 'point-of-view shots' allows the spectator to experience motion from an unseen, performer/camera's perspective. This cannot be achieved in live performance. The spectator is able to watch and empathize with the moving bodies in a stage context, but is not able to 'see/experience' movement through the performer's 'eyes'.

The shot then cuts to a close-up of a spider cupped in a performer's hands, which, in technical terms, is followed by an extremely complicated and lengthy shot. Its complexity derives partly from the fact that the shot employs a considerable degree of camera movement but also that in the finished work the shot is played backwards, hence all of the action had to be planned in reverse. It begins with a long shot of the man with the spider who can be seen in the distance, through a stone window frame. This opening image would have been filmed at the end of the shot. The camera then slowly drops down to reveal a figure lying on a bench, moves along a brick wall, and then on to a medium shot of a performer who appears to be sticking petals onto a rose (in the original shot he was obviously plucking them off). The camera pans along and down to two men who are covered in a blanket of red leaves and rose petals. Although the petals would have originally been dropped onto the men, with the reverse motion, the petals appear to float up into the air away from the men. The shot ends as the camera moves towards the top of a stone staircase where a vial of 'perfume' appears to tip slowly over and seep down the stairs.

A number of points make this shot interesting in terms of video dance. The actual movement is extremely simple; however, the long, unedited sequence and the use of reverse motion create some strong movement images. As all the performers are static, the sense of movement derives more from the slow and sustained camera work that tracks around the whole scenario. Its precise changes of direction also create a clear sense of spatial awareness. It is notable that the movements of the camera construct different shot sizes depending on where the performers are located in relation to it. Even without the use of cuts, the camera is able to pick up on a variety of bodies in close-up, medium distance and long shot. Although the action is minimal, the use of reverse motion alters certain movements. Part of the interest of the sequence is that, as the movement content is subtle, at first it is barely noticeable that the image is being played in reverse. Yet when the performer sticks the rose petals back onto the stalk, his movement appears distorted and surreal. Thus, even with the most simple of technological devices, movement can be manipulated considerably.

The shot then cuts to the white studio. The men lie on a carpet of red leaves and are surrounded by glass shelves filled with 'perfume' bottles in assorted shapes and sizes. The movement content is minimal and consists of the occasional turn of the head or alteration of an arm position. Once again, much of its quality derives from the low-level camera work, which steadily travels around the men, changing from a close-up perspective to a long shot. The image then cuts to a top shot of the men who now appear suspended across the television screen. The pinks of their suits set against the red carpet of leaves create a striking visual image. The movement of the performers continues to be minimal: they raise and lower their heads, roll over and alter the position of their arms. The cuts are occasional and paced evenly, which augments the calm, minimal quality of the dancers' movements. On television, the subtlest of gestures can be imbued with complexity when filmed perceptively within a striking visual context.

An overview of Chapter 3 would suggest that the televisual apparatus is inextricably linked to the construction of the dancing body. In video dance the televisual devices are not minimized or underplayed, but are emphasized, manipulated and experimented with to create cutting-edge dance imagery and unconventional viewing perspectives. The motion of the camera, the rhythm of the edit, the framing of the image, the possibility of special effects, the size of the screen and the quality of the television image all contribute to the type of movement that has emerged in video dance. The implication is that choreographers have had to reconceive how they design dance for television, while existing notions of 'dance' have been called into question by the video dance form. The symbiotic relationship between dance and television marks a 'televisualization' of the dancing body; it marks the emergence of a video dance body with its own set of codes and conventions, which challenges existing representations of dance, both on stage and screen, through an innovative exploration of the televisual apparatus.

4
Postmodern Dance Strategies on Television

Dance, television and postmodernism

In contrast to Chapter 3, which examines how the televisual apparatus acts upon dance, Chapter 4 focuses on the implications of dance in the television context. The evolution of video dance can partly be attributed to developments in film and television, and the significance of the image and technology in late twentieth-century society. Yet the dance context out of which video dance and its practitioners have arisen has also had an impact. The emergence of video dance in the United Kingdom in the 1980s coincides with the beginnings of British postmodern stage dance (Banes, 1987; Mackrell, 1991). The dance artists who became involved in creating video dance works were, and continue to be, primarily located in this particular dance scene. Hence Chapter 4 sets out to examine the significance and implications of dance, and specifically postmodern dance practices, within the television context. Although the evolution, developments and ambiguities of a postmodern stage dance aesthetic are well documented within dance scholarship (Copeland, 1986; Banes, 1987; Briginshaw, 1988; Manning, 1988; Mackrell, 1991; Daly, 1992), the relationship between postmodern dance and postmodernism within other art forms is far from clear. This is highlighted by the plethora of texts that expose the multifarious, and often contradictory, conceptualizations of a postmodern aesthetic within different art practices and cultural formations (Appignanesi, 1986; Kaplan, 1988a; Collins, 1989; Connor, 1989; Hutcheon, 1989; Boyne and Rattansi, 1990; Jameson, 1991; Docherty, 1993).

The implications of postmodern stage dance practices within a television context are dependent on the extent to which television is conceptualized as a postmodern phenomenon. This in itself is a contentious matter. Some of the positions of this argument are revealed in Wyver's (1986) article 'Television and Postmodernism' which attempts to locate a postmodern aesthetic within television. Wyver argues that although television has neither a discernible modernist tradition against which there has been a reaction, nor a recognisable avant-garde, it is possible to identify a type of postmodernist television that is characterized by its challenge to the 'primacy of sight'. This notion is based on the idea that 'seeing is believing' and that, culturally, sight is privileged as 'the truth'. Drawing on the seminal work of McCabe,[1] Wyver suggests that in classic realism there is a hierarchy of discourses in which narrative discourse is dominant. Within film, the narrative is relayed through the camera and, as a result, the camera takes on a position of 'truth'. It could therefore be argued that mainstream narrative cinema supports and perpetuates the primacy of sight (Wyver, 1986). Wyver goes on to suggest that this practice has been echoed in television. Television was originally conceived to be a 'window on the world' that unproblematically relayed real objects and events through to the television screen. In fact, the making of television is very much a technical and political process in that the images are selected, taken out of context and reordered. Yet the belief remains that the television camera neutrally records the events of everyday life. This notion of an unmediated 'truth' has continued with the classic realist conventions of the television drama series. The use of strategies such as realist devices, narrative and character within television, which seek to construct an illusion of truth, has been well documented by Fiske (1989).

In order to differentiate this mode of television, which deals with established literary and dramatic conventions such as narrative and character, from the discussion of a possible postmodern television that follows, the former is conceptualized as a 'classic television framework'. Although this is a somewhat crude division, in that the theoretical parameters of 'classic' and 'postmodern' television are far from discrete, these terms usefully indicate the existence of two distinct theoretical positions. The term 'classic' is employed because, first, it alludes to McCabe's notion of the 'classic realist text';[2] and second, it avoids the complex debate of whether a modernist form of television has existed against which there has been a postmodern reaction.[3] Wyver (1986), along with several other scholars (Connor, 1989; Fiske, 1989; Jameson, 1991; Kellner, 1995; Harris, 1996), has identified examples of television that may be characterized as postmodern. He suggests that a type of postmodern television has emerged that challenges the primacy of sight in that it no longer purports to relay the real. Wyver notes that the defining features of this type of television are its appropriation of images, its borrowing of historical styles, and its effacement of boundaries between high art and popular culture. Within postmodern literature much scholarly attention has been devoted to two areas of television that are said to be characteristically postmodern: pop music video, particularly within the MTV format, and the 1980s 'cop series', *Miami Vice* (Wyver, 1986; Fiske, 1989; Connor, 1989; Kellner, 1995; Harris, 1996). The postmodern features of these texts are the use of pastiche and parody, the recycling of images from other genres, the multiplication and repetition of intertextual images, a liberation of the signifier, a fragmented structure, and a predilection for aesthetic spectacle and surface image.[4]

Although the above examples refer to specific texts, some theorists argue that all television is essentially postmodern in form (Harris, 1996). Connor (1989) usefully summarizes these positions in his analysis of television: whereas high art is characterized in terms of permanence, uniqueness, and authorship, television can be defined by its transience, multiplicity and anonymity. It is marked by an uninterrupted flow of mass produced images. Through the frequent use of the 'remote control' the images are fragmented and eclectic (Connor, 1989). Theorists such as Baudrillard, and Kroker and Cook, conceptualize television as an endless play of random signifiers, to describe the way in which television images have become dislocated from the real; television is constituted by an 'obscene' excess of endlessly proliferating images, which can only result in an implosion and collapse of meaning (Connor, 1989). This conceptualization of postmodern television, however, is somewhat extreme and has been much disputed for its nihilist vision of imploded meanings and ambivalent audiences (Connor, 1989; Kellner, 1995; Harris, 1996). Empirical evidence suggests that contemporary television is not solely constituted from a ceaseless stream of fragmented and meaningless signifiers but continues to employ realist strategies and linear narratives (Fiske, 1989).

It is neither within the scope of this book, nor its purpose, to draw a conclusion as to whether or not television can be characterized within a postmodernist framework. Yet the significance of postmodern dance within the television context is, to some extent, determined by the conceptual model in which television is situated. One of the difficulties in any study of television is that the medium produces a vast number of texts; while some may bear the hallmarks of classic, realist devices, others may display certain postmodern characteristics. It would clearly be impossible to compare and contrast video dance in relation to this considerably wide spectrum of texts. The aim of Chapter 4 is to acknowledge the various theoretical positions that are ascribed to the television context, which range from a classic realist framework through to a postmodern perspective. These theoretical paradigms can provide the foundation for an examination of the relocation of postmodern stage dance practices to the television screen. This then avoids a singular account of the impact of video dance within the television context but recognizes instead the complexities and tensions of this particular investigation, depending on whether television is conceived within a classic or postmodernist framework.

For the purposes of this chapter, four areas of investigation can provide a pertinent discursive structure with which to examine the implications of postmodern dance on screen. They were selected not only because they encompass certain features that are relevant to both dance and television but also because potentially they highlight various tensions and instabilities in the relationship between the practices of postmodern stage dance and the codes and conventions of television. The four areas are as follows: the first deals with the subject of realism and considers this concept in relation to video dance; the second perspective focuses on narrative and how it is affected by postmodern practices; the third area is concerned with the notion of 'character' and how this is set in contrast to the 'performing body' in dance; and the fourth section considers design and aesthetic in video dance and its relationship to postmodernism. As with Chapter 3, Chapter 4 draws on examples of video dance from the *Dance for the Camera* (2 & 3) and *Tights, Camera, Action!* (1 & 2) series in order to illustrate certain concepts. The theoretical ideas that make up this chapter are taken from both film and television theory. In some cases there is a degree of overlap or commonality between film and television theory with regard to certain features of the two media. For example, television theory has appropriated key areas of film theory in relation to the subject of classic realism. At other times, in recognition of the differences between film and television, the two areas of theory take up opposing or divergent theoretical positions. One example is the way in which the two media deal with narrative. In such instances the discrete perspectives of each body of theory are individually stated.

Breaking the realist code

It is suggested in Chapter 3 that the televisual apparatus creates striking and unconventional imagery in video dance as bodies defy gravity, travel at physically impossible speeds and challenge existing perceptions of spatial logic. Video dance has the capacity to disrupt expectations. It refuses to conform to the capabilities and limitations of the live dancing body, but instead transcends them. It is a televisually enhanced body; a body that surpasses reality. The concept of 'reality' is an ideal starting-point to examine the relationship between dance and television as this area of investigation provides an immediate element of tension: at a superficial glance, the two media are seemingly equipped to deal with reality in distinct ways. As television has the technological capacity to make a precise reproduction of objects and events from the 'real world', it is believed to share a close relationship with reality (Wyver, 1986). It is able to construct images that appear to be an almost identical copy of the original subject matter. With dance, however, its subject matter is primarily mediated by the body. Although in some cases the body is used as a figurative representation, in other instances it is employed to deal with completely abstract notions. Therefore, on the surface, it appears that the precise image reproduction of television is more equipped to deal with 're-presenting' reality than dance is.[5] Indeed, this sentiment has been echoed within dance criticism as a means to highlight an opposition between dance and film/television: 'Dance and film are inherently incompatible: film is realistic, dance unrealistic' (Sacks, 1994, p. 24). Yet the links between reality and television are not as straightforward as they may at first seem.

A fundamental debate for film and television scholars centres on the relationship between the image and reality. The complexities of this dialectic are rooted in the ontological questions of whether an objective reality exists, if it can ever be accessed, or whether it will always be mediated by the symbolic structures of language (Fiske and Hartley, 1978; Williams, 1980). Even if some notion of 'the real' is found to exist in the form of the objects and events that constitute the everyday world, the debate continues as to whether the image is a direct reproduction of this reality or whether reality is immediately distorted through the social, cultural and technological operation of the film and television media. Some of the issues of this debate are neatly summarized by Wollen (1969) in his references to Peircean semiotics.

Peirce developed a taxonomy of the sign, which can be divided into three categories: icon, index and symbol (Wollen, 1969). An 'icon' is a sign that represents its object through similarity. This could include a portrait or a diagram. Meanwhile an 'index' 'measures a quality not because it is identical to it but because it has an inherent relationship to it' (Monaco, 1981, p. 133). For instance, a sundial is an index of time, and medical symptoms are indices of health. Finally, a 'symbol' is described as an arbitrary sign that has come to exist through convention. In this instance, there is no resemblance between the sign and its referent; the most obvious example is written language (Wollen, 1969). Peirce states that these categories are not mutually exclusive; yet there has been some debate about how to categorize a photograph,[6] and whether it is iconical or indexical. He proposes that on one level the photographic image purports to be the same as its referent and is therefore iconic; but on the other, as it is physically (or technologically) forced to resemble the referent, the photograph is not identical to its subject matter – yet nor is it arbitrary. Hence the relationship must be indexical. Barthes challenges this stance and posits that the photograph is iconic. Wollen (1969) summarizes Barthes's position in saying...'There is no human intervention, no transformation, no code, between the object and the sign; hence the paradox that a photograph is a message without a code' (p. 124).

Yet Barthes's argument can be disputed. Although a photograph is undoubtedly a technological operation producing images that are closely tied to reality, it still involves an element of human agency in terms of what components are included and omitted, the distance between the camera and the object of focus, whether it is in colour or in black and white, and so on. These very choices constitute a number of photographic codes that construct a particular 're-presentation' of reality. Metz supports this counter-argument in relation to film. He suggests that a photograph is not iconic as an image of a revolver does not simply signify 'revolver'. Its message is far closer to 'here is a revolver', which in turn is a particular colour, shape, weight and type. Similarly, Bazin suggests that a photograph is closer to an index than an icon as it is an 'imprint' of reality rather than reality itself (Wollen, 1969).

Although other related literature suggests that elements of all three categories occur within film (Wollen, 1969; Monaco, 1981; Stam et al., 1992), these few examples usefully highlight the uneasy relationship between the image and reality. This is further problematized by the device of 'realism' which pervades the film and television media. Realism is a set of artistic conventions seeking to convey an impression of reality through specific codes and practices (Wollen, 1969; Willeman, 1972; de Lauretis, 1984; Fiske, 1989). The somewhat tenuous links between realism and reality itself are evident from the range of realist traditions that have evolved. For instance, 'socialist realism' seeks to critique society and offer an alternative existence; 'deep-focus cine-matography' uses 'depth of field'[7] camera work to match the sharp focus range of the human eye; and 'Italian neo-realism' employs non-actors and real-life locations and events to create a greater sense of real-ity (Armes, 1971; Williams, 1980). This diversity of practices suggests that realist conventions are historically, culturally and technologically situated.

McCabe's (1981) concept of the classic realist text is derived from the nineteenth-century realist novel on which, he argues, society's notion of 'the real' has largely been based. McCabe suggests that realism is less to do with authentic subject matter than with the arrangement of dis-courses in the text. This then allows for the possibility of classic realist devices within fantasy genres and improbable narratives. He states: 'A classic realist text may be defined as one in which there is a hierarchy amongst the discourses which compose the text and this hierarchy is defined in terms of an empirical notion of truth' (p. 217).

At the pinnacle of this hierarchy is narrative discourse which McCabe suggests is constructed as a transparent articulation of truth. In the case of the nineteenth-century realist novel, the characters' dis-course is situated within quotation marks and is considered subjective. It is therefore the language outside the speech marks that is seen to be reliable and truthful as it comments on the validity of the characters' speech (McCabe, 1981; Harris, 1996). In film, McCabe suggests that it is the camera which is the dominant discourse within the narration of events. He states: 'The camera shows us what happens – it tells the truth against which we can measure the discourses' (1981, p. 219). McCabe offers two distinct features of the classic realist text which are used later in this section as a means to examine how video dance is situated within this characterization of realism: the first is that the classic realist text cannot deal with the 'real as contradictory'; and the second is that the classic realist text places the subject in a position of 'dominant spec-ularity'. Before investigating these two issues in more detail, it is impor-tant to address another feature of realism: its ideological operation.

It is suggested that, as with film, television also employs realist devices to create a fabricated re-presentation of reality (Wyver, 1986; Fiske, 1989). A tradition has evolved in which 'made-up actors', scenery, costumes, 'eye line match' devices, point of view and 'reverse field' shot structures[8] are used to create an illusion of reality (Willeman, 1972; Fiske, 1989). One of the central controversies of realism is that it is said to have an ideological function as it dupes spectators into a belief that the fictions of film and television are the truth (Comolli and

Narboni, 1971; Martin, 1981). Although realism makes certain omissions and distortions, it nevertheless conveys an impression of reality. This apparent closeness to the real has an ideological effect serving a 'naturalizing process' in which the constructed reality is assumed to be the truth (Fiske and Hartley, 1978; Fiske, 1989; Williams, 1980). Realist practices are said to support the dominant ideology in that they construct hegemonic conceptualizations of reality; it is a 'consumerist, non-critical' way of seeing (Fiske and Hartley, 1978, p. 162) that denies the spectator alternative modes of representation.[9] The question is where postmodern dance practices within video dance are situated in relation to these ideas.

Metz (1975) propounds that in fiction film (and in television texts that employ realist devices) the role of the cinematic (or televisual) signifiers is to erase their presence in order to give the spectator an illusion of reality. That is to say, the various cinematic and televisual codes must be made 'invisible' so that an impression of unmediated reality is constructed. In contrast, one of the characteristics of postmodern dance is a concern with exposing and playing on its own conventions (Mackrell, 1991; Siegel, 1992). This deconstructive strategy can be seen in numerous video dance pieces that draw attention to, and subvert, the televisual codes. One example of this is in *Codex*. In the first section, the movement is speeded up, it is played in reverse, the image is turned upside down, and top shots are employed in order to construct eye-catching images. A similar example is in the second section of *Codex* where a man dressed in a neck brace and rubber dungarees sings a haunting melody while two male dancers perform a contact duet several feet away. At one moment, the camera is positioned so that the singer's facial profile is captured in a tight close-up, while one of the dancers is lying down in the background. Due to the two-dimensional screen, an enormous head appears to be situated next to a minute body. Thus it could be suggested that *Codex* exposes and manipulates the televisual apparatus in order to reveal its own 'constructedness'.

A similar example is with *Monologue*, which employs a single, static close-up of the performer's face as she addresses the camera. One of the realist conventions of film and television is that the camera switches between different shot sizes or positions as this is believed to construct a greater illusion of reality (Monaco, 1981). As Lockyer (1983) notes, this is somewhat similar to the way in which the human subject alternates between looking at the precise detail of objects and events and then looking at them with a more panoramic view. Hence in *Monologue*, as the dancer grows increasingly frustrated and angry, the spectator becomes very aware of the persistent claustrophobia of the unedited close-up. This is further augmented by a series of other breaks with realist televisual conventions: the spectator is given no contextual information to suggest the whereabouts of the performer and the motivation for her actions; she shouts in a strong dialect of vernacular French so that the language itself is almost incomprehensible; she appears to wear no make-up, her hair is messy and she drools as she shouts. Although the 'talking head' is a ubiquitous image within television texts (Fiske, 1989), in these particular instances the face is well coiffured and made up, its verbal message is informative and coherent, and it is intercut with other images of events and bodies which the 'talking head' refers to or communicates with. The absence of cuts, the black-and-white film, and the lack of logic and rationale in *Monologue* has the potential to denaturalize the realist illusion and expose televisual practices as a set of conventions. It is perhaps not surprising that the codes and conventions of television are called into question in video dance, given that the aim of the genre is an innovative exploration of the relationship between dance and television.

It could also be suggested that *Monologue* has a feminist agenda. One of the features that has come to characterize certain postmodern dance practices is a dismantling of hegemonic social structures, and a body of practitioners has emerged who employ feminist deconstructive strategies in order to expose the constructions of femininity (Mackrell, 1991; Brown, 1994). From the 'glamour shots' of female film stars (Kuhn, 1985) through to the attractive 'talking heads' of women presenters, the close-up image of the female face is ubiquitous in film and television. Although *Monologue* uses the close-up convention, through the distraught, homely face of the performer and her stinging, verbal attack, it denaturalizes the image of 'woman' as attractive and glamorous.

Wyver (1986) argues that one way in which the classic realist text can be called into question is through a challenge to the primacy of sight. This can be seen in television texts that make no pretence to be 'relaying the real'. For instance, in the video dance piece *Outside In* (choreographer: Victoria Marks, director: Margaret Williams, 1995) a series of peculiar and unconnected events occur. Although the piece employs the same performers, their various actions appear to have no logic or connection: they pass a breath of air to one another, execute a tango-based sequence, appear under a square of turf, make paint tracks with the tyres on their wheelchairs and walk across a vast green hill. This seemingly unconnected passage of events is typical of the fragmented narratives of postmodern dance forms (Banes, 1987; Foster, 1992b).

Kissy Suzuki Suck turns to prostitution for its subject matter and employs a collage style construction. The film is made up of several distinct discourses: the spectator sees two women smoking and chewing gum in a state of boredom; at one point they execute a series of obscene, sexual gestures; a grainy, black-and-white image that is similar to the 'authoritative' pictures of surveillance cameras is employed; the film is shot with a shaky, hand-held camera style, which is typical of home movies and certain styles of documentary film making; an interview with a prostitute forms part of the soundtrack; later a monologue of 'dirty talk' can be heard; and towards the end, the image switches to bright, synthetic colours as the shape of a woman writhes across the bonnet of a car. Drawing on McCabe's (1981) notion of a classic realist text, *Kissy Suzuki Suck* challenges this concept as there is no hierarchy of discourses. The images depict numerous perspectives on prostitution and the collage-like style refuses to highlight a dominant narrative. As McCabe notes, the classic realist text is unable to deal with the 'real as contradictory'. In contrast, *Kissy Suzuki Suck* simultaneously shows the disapproving morality of the surveillance camera alongside the two women who appear casual and indifferent to their profession, and in doing so refuses to privilege a particular 'truth'.

It is suggested that through the conventions of 'invisible editing' and 'point-of-view' shot structures, the spectator is said to be 'sutured'[10] into the text (Fiske, 1989). This effect is applicable to both factual and fictional programming and places the spectator within a position of 'dominant specularity':

> by allowing the spectator to see and understand the action from (nearly) the points of view of the characters it maintains the impression that the screen gives us direct unmediated access to the action, and thus allows the filmic/televisual metadiscourse to remain invisible, unspoken. Fiske (1989, p. 28)

In the case of *Kissy, Suzuki, Suck,* however, the grainy image, the hand-held camera style and the use of a collage structure offer multiple perspectives on the theme of prostitution, and thus refuse a position of dominant specularity for the spectator. These various formal properties are again typical of a postmodern dance aesthetic. The construction of multiple viewpoints and the abandonment of traditional hierarchies are characteristic of postmodern stage dance practices (Banes, 1987; Mackrell, 1991).

There are of course certain realist genres that also expose and subvert the codes and conventions of film and television (Fiske, 1989; Williams,

1980). For instance, unlike the slick presentation and seamless editing of fictional realism, poor-quality images, shaky camera work and glaring cuts are features of some styles of documentary realism. Yet although video dance regularly highlights the televisual apparatus through which it is constructed, it serves a quite different function from the documentary and, indeed, other forms of realism. The documentary is a vehicle for the explicit presentation of 'factual information' (Armes, 1971), as with all genres of realism, that through a number of devices attempts to create an impression of reality. Video dance, on the other hand, has no such agenda. The genre is specifically concerned with an aesthetic function and has no commitment to 'relay', or construct an illusion of, the real. That is not to suggest that video dance is completely opposed to realist strategies. *T-Dance* is one example of a classic realist text. The piece is based on a senior citizen tea-dance, and the work employs a linear narrative and coherent shot structures in order to offer[11] the spectator a position of dominant specularity. The opening sections of *Alistair Fish* and *Drip* also employ classic realist devices, which is perhaps another approach to exploring televisual conventions in relation to dance; this type of experimentation is a fundamental aim of video dance. These examples, however, are very much in the minority.

The resistance to realist strategies in video dance appears to be dependent on several factors. First, although there is a tradition of narrativity in Western theatre dance, there is also a well-established practice of abstract or formalist dance.[12] This is perhaps one of the ways in which video dance plays a distinctive or significant role in television. Whereas it is commonplace for dance to deal solely with formal concerns, television rarely sets out purely to explore its own form (Rubidge, 1984). Hence, dance has the potential to bring a new aesthetic to television. A second possibility is that part of the focus of video dance has been to explore the creative relationship between dance and television. With this emphasis on innovation, it is hardly surprising that television codes are exposed and manipulated to challenge existing television conventions such as realism. Finally, this challenge to realist conventions may also be due to the postmodern stage context out of which video dance has emerged: the deconstruction of hierarchical frameworks, the play on conventions and the subversion of existing apparatuses are prominent features of the postmodern dance aesthetic. It could be suggested that video dance subverts realist devices and their potential for ideological positioning, although it is not alone in this practice. Harris (1996) identifies a postmodern aesthetic in certain film and television texts that is no longer concerned with realist representation and narrative. It is perhaps not surprising that video dance can be located within this postmodern aesthetic in that it derives from a postmodern stage dance tradition. What is significant for video dance is its capacity to destabilize norms.

Fragmented narratives and episodic structures

As video dance has a tendency to resist realist practices, this results in unusual and unexpected images and multiple perspectives on a theme. This element of disruption is not only apparent in terms of the content of images but also in the way in which video dance is structured. It would seem that the presence of postmodern stage dance strategies in video dance clearly has the potential to influence the order of shots. Therefore this current section investigates how the events of video dance are arranged through time, and how this compares to and contrasts with dance and televisual practices. The sequential arrangement of any events through time has been referred to as 'syntagmatic structure' (Stam et al., 1992). Much has been written on the syntagmatic structures of film and television (Metz, 1974; Monaco, 1981; Silverstone, 1981; Fiske, 1989; Stam et al., 1992) and is addressed under the rubric of 'narrative'. This is because narrative is the predominant structuring device within the film and television media. Hence, before addressing the syntagmatic structure of video dance, it would be useful to examine how the 'narrative form' has been dealt with in film and television.

Fiske (1989) suggests that the narrative model is a universal and ubiquitous structuring device that shares certain characteristics with language. He states:

> Like language, narrative is a basic way of making sense of our experience of the real, and structuralists have argued that it shares many of language's properties, that it is structured along the twin axes of the paradigmatic and syntagmatic. (p. 128)

Film narrative is said to operate through these two dimensions: the syntagmatic structure refers to the arrangement of a passage of events through time, and the paradigmatic dimension is concerned with non-temporal factors such as shot size, lighting, character and setting (Fiske, 1989; Stam et al., 1992).[13]

The characterization of narrative as the temporal organization of events is supported by Stam et al.(1992), who describe narrative as 'the recounting of two or more events (or a situation and an event) that are logically connected, occur over time, and are linked by a consistent subject into a whole' (p. 69).

A vast body of work has developed within film and television theory that addresses the way in which narrative is structured and operates (Metz, 1974; Monaco, 1981; Silverstone, 1981; Fiske, 1989; Stam et al., 1992). Although in some instances these studies have been based on narrative forms within other disciplines, such as literature and structural anthropology, the following examples outline the key narrative theories that have been applied to film and television scholarship: Russian formalism's analysis of the 'fabula' and 'syuzhet';[14] Propp's morphology of the folk tale; Levi-Strauss's study of binary oppositions within mythic texts; Barthes's typology of narrative codes; and Metz's development of the 'grand syntagmatique'. What characterizes these studies is an attempt to extract universal structuring devices applicable to all narrative forms.

It is suggested in the previous section that classic realist texts pervade the film and television media and one of the features of the classic realist text is the way in which the passage of events is organized. Although this theory evolved within film studies, it may equally be applied to television. Fiske (1989) states that the 'classic realist narrative and its preferred reading strategy try to construct a self-contained, internally consistent world which is real-seeming' (p. 130). The classic narrative is characterized through certain formal features: an equilibrium is broken that must be resolved; 'cause and effect' strategies move the narrative forward; spatial and temporal coherence is employed; the characters are psychologically motivated; and the form is linear (Fiske, 1989; Stam et al., 1992). These characteristics are achieved through specific filmic and televisual conventions such as invisible editing, point-of-view shot structures and devices that suggest the passage of time such as 'fade outs' and 'dissolves'. Yet as with classic realism, the classic narrative carries ideological implications. Fiske (1989) argues that the classic realist narrative has an illusory relationship with the real. It gives a sense of coherence and resolution to a reality that in actual fact has neither and excludes anything that may contradict or destabilize this representation. The classic realist narrative is simply a set of conventions. Stam et al. (1992) support this argument through stating that realist narrative devices serve a naturalizing process; by effacing the operation of the filmic (and televisual) apparatus, this constructs a purportedly transparent image of reality. This is said to position the spectator in a place of 'dominant specularity' (Fiske, 1989; Stam et al., 1992), although the spectator may of course opt to resist this positioning (Harris, 1996).

It has been suggested that narrative forms are based on the psychoanalytic structures of castration disavowal and the Oedipal scenario (Kaplan, 1988a; Johnston, 1990; Stacey, 1992). This is tied up with questions of origin; certain narrative clues must be answered and truths resolved in order for the agent in the Oedipal scenario, or for the reader/protagonist in the text, to discover a new equilibrium. Johnston (1990) states:

> The child's refusal to recognize the reality of such a traumatic perception (the absence of the woman's penis) can be likened to that of the reader/viewer, who knows these are just words/images, and yet derives his/her pleasure and security from following the narrative through to the end. (p. 64)

Yet these ideas are based on theory which manifests a problematic epistemology. This is in reference to the phallocentric perspective in which the male is placed as a central marker against which the female represents a 'lack'. Although notions of castration disavowal can be legitimately applied to the male, on the basis of his fear that he might lose his penis, this theory appears redundant when applied to women: it is surely strange that they would fear the loss of something that they never possessed in the first place. It is also notable that the agents who typically motivate the narrative are male and it is the male child who is at the centre of the classic Oedipal complex (Stacey, 1992).

Yet, irrespective of psychoanalytic theory, Mercer (1981) posits that narrative forms are closely tied to concepts of pleasure and desire. Through the way in which the hermeneutic code operates, the spectator has a desire to resolve clues and find a 'truth'. Narrative therefore creates an intimacy with the spectator who is constructed as a desiring subject. In narrative fictions, the way in which the text is structured causes the spectator to identify with certain characters. Yet this process is said to have ideological implications in that identification is based on dominant cultural frameworks (Stam et al., 1992; Burt, 1995)[15] For instance, feminist film theorists argue that representations of gender within mainstream, Hollywood cinema are perpetually constructed within stereotypical imagery (Kuhn, 1985; Mulvey, 1989; Kaplan, 1990;

Creed, 1993). For this reason feminist filmmakers have attempted to employ anti-narrative and non-realist devices as a challenge to the dominance of the classic, realist form and its structures of character identification. Mulvey (1989) describes the destruction of pleasure as a 'radical weapon' against the dominant, Hollywood ideology.[16]

The classic realist narrative is predominant in mainstream film (Stam et al., 1992) and, although the narrative form equally pervades the television context (Fiske, 1989), it operates in a different way from filmic narrative. Within television, the narrative form is immediately recognizable in such genres as drama serials and soap operas, but it is also manifest in other forms such as news programmes and quiz shows. Although these latter examples do not employ classic narrative systems with psychologically motivated characters and explicit plot structures, both genres are based on narrative events, whether through the mini-narratives that are recounted by the news reader or the contestant who must overcome various tasks in order to win the quiz (Fiske, 1989). Unlike film, television narratives are said to be dominated by the serial or series format[17] and, consequently, never reach a point of closure: although the syntagmatic chain of events may be resolved within an episode, the paradigmatic dimension of character and setting remains open ended (Fiske, 1989). Television narratives can be highly fragmented. This is partly due to the commercial breaks, news items, trailers and presenter announcements that regularly interrupt the narratives of television (Fiske, 1989), but it is also symptomatic of the fragmented viewing patterns that characterize television watching and the possibility of switching back and forth across channels with the remote control (Connor, 1989; Kellner, 1995; Morley, 1995). This conceptualization of television texts as fragmented and eclectic is typical of postmodern definitions of television (Connor, 1989; Kellner, 1995; Harris, 1996); yet Fiske (1989) insists that, although television displays significant differences from realism within film and literature, the realist narrative is its dominant form of representation.

There is considerable evidence to suggest that video dance operates outside classic narrative structures. It is suggested earlier that narrative is characterized as a passage of two or more events through time (or an event and a situation) that are logically linked into a whole. One approach within film studies to narrative construction has been to examine the arrangement of shot structures (Metz, 1974; Stam et al., 1992). There are several video dance examples that revolve around a single event, and which lack the linear and logical progression and the establishment of a new equilibrium that are characteristic of classic narrative systems. Indeed, both *Monologue* and *Hands* occur within the space of a single shot and a single event; there is no logical or linear progression through time and no sense of 'new whole' or 'resolved closure'. *Hands* consists of a set of formal hand gestures, although there is no discernible logic to the order of movement. Similarly, in *Monologue*, the hysterical outburst of the performer is distorted through the use of a regional French dialect so that it is impossible to extract a sense of linear narrativity in her speech. An element of narrative resolution perhaps occurs when she stops shouting, but questions regarding the identity of the woman, her location and the reasons for her diatribe remain unanswered.

Although *Le P'tit Bal Perdu* and *Bruce McLean* employ numerous shots, both are based within the context of a single event and avoid linearity or spatio-temporal coherence. In *Le P'tit Bal Perdu*, two performers sit at a table in the middle of a vast green field and execute a complex sequence of gestures. The organization of shots and structure of movement revolve around the arrangement of the accompanying prerecorded song. During the chorus, certain shot structures and movement sequences are identically repeated, and with purely musical sections the shot cuts away to a herd of cows and an accordion player situated elsewhere in the field. The close relationship between the sequence of movement and the choice of shots, in relation to the structure of the song, resist the linearity that is typical of classic narratives. A similar example is with the *Bruce McLean* video. The piece is based on a middle-aged man in a studio who performs a series of pedestrian activities, but as they are rapidly edited and reordered into a fragmented and repetitive structure, this results in a disruption of spatial and temporal coherence. There is no sense of pattern or logic to the sequence of his action.

This fragmented and episodic arrangement is typical of video dance. *KOK* draws on imagery from a boxing match, but there is no logical structure or sense of chronology to the relay of events. At one point a close-up is depicted of four people, face-on to the camera; in unison, they pant, blow and duck back and forth. This is followed by a series of boxing images in which the dancers pose in stillness. The piece continues with other recognisable images from a boxing match, but there is no narrative in the sense of a competition that is then won or lost. The shots are fragmented and disordered. Likewise, *Perfect Moment* consists of a number of events arranged in an episodic and aleatory form: a group of people, dressed in white bath robes, act out cosmetic activities; four men, in vests and kilts, stride, hop and skip across a courtyard;

various people, wearing wetsuits, execute a quirky movement sequence on a bed of sand; and women pose around in extravagant and glamorous ball gowns. Except for the fact that it is the same group of performers throughout the work, there is no logic or continuity to the sequence of events.[18]

Several video dance pieces contain an element of narrativity; yet they do not necessarily conform to the characteristics of the classic realist narrative. *Alistair Fish* follows the eponymous hero who finishes a swimming session to discover a note from his girlfriend stating that she has left him. He sets off on a train to pursue her, but it is at this stage that the classic narrative becomes suspended. The scenarios that follow are bizarre and episodic. There is no longer any obvious connection between the various occurrences and the motivation of the narrative, as curious events develop between Alistair and the other passengers: they nod off to sleep in unison; they dance through the aisles of the carriage; a small boy fires arrows at Alistair and later tap dances on the table; bodies lie in luggage racks; and various passengers chase him through the train. The events are strange and incoherent and the characters' actions are not logically motivated. Peculiar and episodic narrative structures can also be seen in *Drip* and *The Storm*. *Drip* consists of a somewhat strained birthday party that includes a Latin American dance duo who burst onto the scene and a homoerotic fantasy between a dull, anorak-clad guest and a 'Valentino-styled' narcissist, obsessed with taking Polaroids of his various body parts. *The Storm* features a wild contact duet between a drenched young couple in the midst of a smart and sedate restaurant.

Within the *Dance for the Camera 2 & 3* and *Tights, Camera, Action! 1 & 2* series, only three works employ classic, realist narratives: *T-Dance* follows the events of a senior citizen tea-dance; *Never Say Die* (choreographer/director: Nigel Charnock, 1996) is based on a bitter and twisted ballet teacher who, in a fit of jealousy, shoots her talented young protégée; and *L'Envol de Lilith* is styled as a black-and-white silent film and follows the fortunes of Lilith, an exotic flamenco dancer. All three works employ linear narratives, cause and effect action, spatial and temporal coherence, and psychologically motivated characters; however, *Never Say Die* and *L'Envol de Lilith* perhaps resist the three-dimensional characterization of classic realism in their use of stereotypical characters. *Never Say Die* features the cliché of the mature dancer who, resentful over her lost youth and playing on another balletic cliché that links dance with eating disorders, is caught vomiting down the lavatory. *L'Envol de Lilith* is a postmodern pastiche that draws upon

Orientalist imagery and the stock characters of a hot-blooded matador, a dark, silent caliph and the sultry Lilith.

The absence of classic realist narratives in video dance, and the narrative form in general, is due to several factors. As stated in the previous section, there is a well-established tradition of formal or pure dance within the theatre dance context and it is perhaps not surprising that, in some instances, video dance continues to employ this tradition. Video dance works such as *Attitude* and *Pace* are solely based on formal dance concerns: *Attitude* is a blend of Graham-derived contemporary and jazz dance, shot in the smoky haze of a blues club; and *Pace* is a rapidly edited film of a performer who executes a release-based solo. The influence of postmodern stage dance practices can also be seen in the syntagmatic structure of certain video dance pieces, as eclectic, repetitive and fragmented narratives are characteristic features of postmodern dance (Banes, 1987; Foster, 1992b). Indeed, it has been suggested that these postmodern devices are more appropriate for the television medium. Whereas ballet is structured towards a climax, postmodern dance is made up of 'equally valid moments' (Brooks, 1993, p. 24). This is more suited to the television medium, which demands radical changes of image and regular moments of climax in order to keep the spectator interested (Rubidge, 1988b). Although these postmodern devices complement the medium, the absence of narrative in video dance is perhaps a challenge to the conventional television aesthetic. Again, it is possibly video dance's commitment to explore the relations between dance and television that has prompted the resistance to, and subversion of, existing traditions.

As suggested at the beginning of the chapter, the significance of the syntagmatic structure of video dance in relation to the television context is dependent on whether television is conceptualized within a classic or postmodernist framework. Fiske (1989) argues that television is dominated by narrative forms, thus it could be suggested that video dance is somewhat subversive in its resistance to narrative structures for it avoids the binary oppositions of psychoanalytic structures and linguistic formations that have been attributed to narrative (Fiske, 1989; Stam et al., 1992). From a postmodernist perspective, however, video dance is typical of the eclectic, repetitive and fragmented narratives that characterize television viewing practices and the flow of images on screen. In order to examine further the significance of dance, and specifically postmodern stage dance practices, in the television context, the following section deals with the notion of a 'performing body'.

The performing body

The 'subjects' or 'players' of television constitute a diverse spectrum of individuals, ranging from the actual 'dancers' of video dance through to the actors, characters, presenters, everyday people, stars and celebrities who make up other areas of the television network. The notion of a 'performing body', however, can act as a useful umbrella term with which to encompass all of these diverse subject types. The idea of a 'performance' highlights an element of staging or constructedness. This would then suggest that the players of television are not simply 'real' or 'unmediated' bodies, but are in fact constructed and framed by the televisual apparatus. The concept of a 'performing body' is applicable to all the players of television from the 'video dancers' themselves, through to the everyday people of 'fly-on-the-wall' documentaries. Thus the focus of this section is to examine how the 'performing bodies' of television are conceptualized within related academic literature and where video dance, particularly under the influence of postmodern stage dance practices, stands in relation to these constructs.

It is suggested earlier in the chapter that the film and television media are dominated by realist practices. Consequently, psychologically motivated characters, whose actions further the narrative, constitute a central component of the classic realist text (Cook, 1985; Fiske, 1989). The 'character' is a dominant feature of television realism. A character is a constructed representation of a person: a fictional being whose actions advance the plot structure (Dyer, 1981). Although each character is obviously played by an actor, the character is inscribed with realistic discourses, which makes it difficult to separate the two. Characters are seen to be realistic as they display an element of psychological complexity, their actions are motivated, they develop through time, they appear to have 'lives of their own' outside the plot, and although, in some instances, their actions may be unexpected, these conform to the broad parameters of the character's personality traits (Dyer, 1981; Fiske, 1989).

The practice of characterization serves a naturalizing process in that, as a result of realist devices, a fictional construct appears as to be a 'reality'. Fiske (1989) argues that this naturalizing process is largely to do with the sense of uniqueness or individuality that a character portrays. This is based on the bourgeois notion of the individual as an 'agent' whose personal conscience can cause historical and social change (Dyer, 1981). The bourgeois subject is conceptualized as a 'natural' phenomenon, which conveniently manoeuvres attention away from notions of social construction in relation to subject formation. As Fiske (1989) states:

> Like realism, individualism sees the self as the prime site for unify-ing and making sense of experience; a unified sense of experience produces and is produced by a unified sense of the self. This self is unique and different from other selves: its origins are rarely exam-ined but are assumed to be natural, or biological, rather than social. (p. 152)

Although the character is a feature of both the film and television media, there is a fundamental difference between the two. Whereas film is conceptualized as a recounting of an existing narrative, televi-sion is characterized by an illusion of 'nowness' or 'liveness' (Fiske, 1989). The emergence of the serial or series format in television allows its characters to be presented on a daily or weekly basis, which adds to the sense of immediacy, but in film the characters tend to feature only in the recalling of one particular passage of events.[19] In the television series the characters do not change between episodes, whereas in the television serial they continue to live or develop between and during each episode. Yet regardless of these specificities, the employment of character as a filmic and televisual convention carries with it certain implications.

Various identification theories which address the way in which the spectator relates to fictional characters have evolved within film and television studies. For instance, one strand of film theory posits that cinematic identification derives from the subject's early psychic devel-opments as a child and this is located in the Lacanian 'mirror stage' (Metz, 1975; Friedberg, 1990; Stam et al., 1992). Prior to this period the child perceives itself only as a shapeless, fragmented body, but at around the age of six to eighteen months, the child sees its reflection in a mirror and identifies with this unified shape as being a 'superior self' (Sarup, 1988; Stam et al., 1992). Later the child acknowledges that the reflection is a separate entity, but this recognition of the 'other' forms the child's sense of self through the 'I–you' differentiation. It is for this reason that throughout life the subject constantly attempts to identify with the 'other'. The film theorist Metz (1975) writes that cinematic identification is similar to the mirror stage, in that the screen is like the mirror in which the spectator seeks to identify with the 'image as other'. The spectator is able to accept her or his absence from the screen as she or he has already come to terms with the initial mirror stage. Metz suggests that primary cinematic identification is with the 'act of looking',[20] but it is secondary cinematic identification that is most appropriate to the notion of character. He asserts that the spectator identifies with certain camera positions and shot structures that relate events from a character's (or several characters') point of view. Although, as previously discussed, there is a need to be cautious when dealing with a psychoanalytic epistemology as a theoretical tool, Metz's conceptualization of the way in which the spectator is positioned through the filmic apparatus is an ideal vehicle through which to expose the ideological agenda of film. Through realist strategies, film creates an illusion of ideological neutrality even though, through certain shot structures, the spectator sees events from a privileged position which, in mainstream film, is said to support the dominant ideology (Kaplan, 1990; Stam et al., 1992).[21]

As television theory has been particularly concerned with the social nature of its audience rather than with the homogeneous spectatorship position that has characterized much film theory, several studies have been made into the way in which the television spectator 'reads' the medium (Fiske, 1989; Morley, 1981; Harris, 1996). Fiske (1989) argues that although audience members often predict and review the actions of a character, or refer to and communicate with actors through the names of their characters, this does not automatically mean that the spectator is unable to distinguish between representation and reality. He suggests instead that the viewer is simultaneously involved with and detached from the character. Fiske describes this as 'implication', which constitutes part of the pleasure of television viewing; rather than a type of 'passive identification', it implies that the spectator asserts a willingness to play along with the fantasy (Fiske, 1989).

Both film and television theorists have nevertheless commented that characters are predominantly constructed through hegemonic discourses (Kuhn, 1985; Fiske, 1989; Dyer, 1993). This would then suggest that, irrespective of how closely a spectator identifies with a given character, there is always a danger that the character is embedded in, and maintains, the dominant ideology, as there are a lack of alternative representations. Fiske (1989) does present an oppositional approach in the form of discursive readings. He suggests that a structuralist analysis conceptualizes the character as a signifying system that embodies certain social and economic values. Fiske goes on to assert that this type of reading can activate contradictions, which allow for dominant, alternative and resistant readings. It would seem, however, that with Fiske's (1989) notion of a discursive reading the spectator is rooted in an ideological impasse. The spectator can either accept or resist the dominant ideology, but the possibility of transgression does not appear to be an option.

With some of these ideas in mind, it would now be useful to begin to question where video dance lies in relation to character and characterization. With the exception of *T-Dance*, the majority of video dance examples reveal an absence of classic realist narratives that employ 'rounded' and psychologically motivated characters. This may largely be because video dance draws on a dance tradition that is composed through a different set of codes and conventions from the practices of television. The following examples examine how video dance deals with the performing body and the extent to which this may reflect various dance practices, particularly from the postmodern genre.

In keeping with the pure dance tradition that constitutes some forms of Western theatre dance, the first section of *Codex* is a prime example of the performing body as an 'abstract representation'. In this particular section the dancers are completely covered in skin-tight, black body suits; they wear gloves, hoods that mask their faces, and diving flippers on their feet. The result is that sex, age and ethnicity are masked and, consequently, the spectator can neither differentiate one performer from another nor begin to categorize them as 'character types'. This abstraction and homogenization is further augmented by the movement material and shot structures. The dance content is predominantly formalist and is based on tilts and upper-back curves, running, small jumps, diagonal arm positions, *demi-pliés* and tapping feet. The group of performers shares the same movement material so that no one is made to seem distinct, except perhaps for the final image of the body executing a series of hand springs across the floor (although the spectator cannot tell which performer this actually is). The camera work meanwhile remains face-on to the action so that the identical bodies are rendered as abstract patterns that mutate across the screen. There are no point of view shot structures and, even if there were, they would fail to be effective as the spectator is unable to distinguish one performer from another. Although this representation of the body as a formal entity is unconventional in televisual terms, the use of abstraction is a common device within Western theatre dance. This practice can also be seen in *Echo* (choreography: Mark Baldwin, director: Ross MacGibbon, 1996), a formal piece of choreography designed in relation to the sculptures of Anish Kapoor, and *Keshioko*, which draws on pure dance concerns and is set within the vast industrial landscape of a dockyard.

Not all video dance, however, employs such abstract modes of representation. For instance, *Tango* uses a number of recognizable 'types' who perform everyday actions. The piece is set in a room and, through a process of animation and accumulation, various bodies repeatedly enter and exit the room and, on each occasion, execute an identical set of actions: a small boy retrieves a red ball which has bounced through the open window; a robber sneaks in to collect a package; a mother feeds her baby; a jogger does a handstand on the table, and so on. Yet although the performers are clearly signified as recognizable 'types', through such codes as costume, action, sex and age, the repetition of action and absence of interaction between the bodies suggests that they do not possess the psychological motivation or development through time that constitute a character. It appears that the key factor behind the work is an exploration of formalist concerns rather than any discursive or representational aims. Once again, this is typical of the well-established formal tradition that has come to characterize certain areas of Western theatre dance.

La La La Duo No 1 consists of a vigorous contact duet between a man and woman. A diversity of representations is clearly inscribed in the choreography, which results in certain moods and emotions emanating from the piece. For example, at one moment images of a ballerina come to the fore as the woman, who wears a black tutu, is lifted in a 'fish dive'; at another moment her flirtatious poses make her out to be a model or a film star. In one instance, she resembles a female weightlifter as she staggers forward while carrying her partner above her head; yet moments later she is vulnerable and child-like as she squats on the floor. Although there are fleeting references to figurative representations, they are neither stable nor fixed. This multiplicity and instability of representation is a common device within postmodern stage dance: the spectator is not presented with a single and coherent view of the world but a multitude of contradictory references (Banes, 1987; Mackrell, 1991). This is typical of the fragmented and nomadic identity ascribed to the postmodern subject (Kellner, 1995; Harris, 1996). To some extent, the instability of the performing body in *La La La Duo No 1* can be seen as a subversive device in relation to the relative stasis of conventional characterization within the film and television media. This tendency for the body within video dance to avoid the naturalizing processes of realist characters may also be read as a resistance to dominant ideological forces.

The sense of fragmentation that characterizes *La La La Duo No 1* can also be seen in *boy* and *Horseplay*. Both pieces draw on recognizable 'character types', but there is no sense of development of character through time or psychologically motivated action that furthers the narrative. In both instances the syntagmatic structure consists of episodic fragments that have no logical or linear order. In *boy* the young child races like a wild cat along the coastline, surveys his territory as if in preparation for combat, leaps over the dunes with a carefree abandon and executes a movement signature, combining martial arts with secret spell making. Similarly, *Horseplay* is structured through fragments of action that involve three young women; they blow bubble gum in each other's faces, flirt with a young man and try out various 'street dance' moves. In both examples certain character types emerge, but their relation to the action is fragmented and episodic rather than linear and motivated.

Another feature of televisual representation is in the form of 'everyday people' or people who purport to play themselves (Fiske, 1989). For instance, the television spectator often sees images of 'everyday people' who just happen to be captured on location shots, or are filmed for documentary or news purposes. A similar example is with television presenters such as news readers or quiz show hosts who appear to be themselves. It has been suggested, however, that this is simply another ideological facet of realism that constructs an impression that these bodies are 'real', as they are situated outside explicit fictional narratives. Yet these people are as equally constructed and codified as fictional characters. For example, news readers, quiz show hosts and everyday people on the street are framed within certain shot structures, filming devices, and specific styles of speech and dress (Fiske and Hartley, 1978; Fiske, 1989; Morley, 1981).

Several examples of video dance use 'everyday people', many of whom have had no formal dance training. For instance, in both *Mothers and Daughters* and *Man Act* the spectator is informed that these are, respectively, 'real-life' mothers and daughters, and fathers and sons, and *Relatives* is an autobiographical piece in which Ishmael Huston-Jones improvises to an unrehearsed monologue by his mother. The implication is that these are 'authentic' familial relationships; yet irrespective of whether or not this is the 'truth', since video dance is not bound by realist conventions, the validity of these claims is unimportant. Whereas the 'everyday people' of television texts are situated within realist practices, the 'everyday people' of video dance are placed within certain spatial, temporal and dynamic structures that remove the work from a sense of verisimilitude. For instance, in typical postmodern fashion, *Relatives* employs a collage-style structure and *Mother and Daughters* presents a multiplicity of perspectives on this theme. Through its subversion of realist conventions, video dance perhaps avoids the ideological illusions of conventional television realism.

The final theory of the performing body that needs to be briefly addressed in relation to film and television is the concept of 'celebrity' or 'star'. These terms refer to famous individuals who appear on film and television and, although they carry slightly different connotations, they share similar functions in relation to reading practices. Whereas the term 'celebrity' or 'personality' is conventionally used to refer to well-known television people, the term 'star' has generally been associated with people in the film industry. The other significant difference is that while 'stars' are signified as unattainable and distanced from their fans, 'celebrities' are more familiar and accessible figures (Fiske, 1989). Yet both categories involve reading practices that are tied up with intertextual references. Dyer, 1981 (Fiske, 1989; Turner, 1993) posits that a 'star' is a signifying system which carries various connotations independently of the character that she or he may be playing:

> the construction of a star's image within the culture incorporates representations of him or her across an enormous range of media and contexts; fan magazines as well as the mainstream press, television items and profiles as well as performances in films, conversations and heresay as well as personal appearances. Turner (1993, p. 103)

Consequently the spectator is said to read these intertextual references into the various performances that a 'star' may give. The same can be said of the television celebrity and, as with film, it often becomes difficult to separate the star or celebrity from the actual character. Yet the notions of star and celebrity are not particularly pertinent to video dance in the sense that, except for a handful of well-known ballet stars, the performers in video dance are not household names. Only an extremely specialist dance audience would be able to read any intertextual references into the work of a particular performer. This perhaps suggests that the 'players' in video dance possess more 'fluid' and nomadic performing bodies than the limited associations and representations of the star or celebrity. The relatively fixed persona of the star or celebrity is in keeping with the bourgeois notion of a fixed and unique identity, whereas the performers of video dance are much closer to the postmodern notion of a mobile and fragmented identity. This instability of representation is a potentially powerful position as it resists the stasis and fixity of bourgeois notions of identity.

Seizing the spectator's eye

With its bold, playful and unpredictable images, video dance calls into question our expectations of the television medium. In the same way that it challenges notions of realism, narrative and character, the manipulation of design and aesthetic in video dance disrupts existing conventions. In order to explore this area in more detail a consideration of the viewing conditions that characterize television can provide a useful starting-point. Although television tends to be seen as a visual medium, it is primarily used as an aural one (Lockyer, 1993; Morley, 1995). This is largely to do with its viewing context which differs considerably from film viewing. Whereas cinema is a communal activity that involves travelling to the venue, buying a ticket and watching the film in a darkened room during one continuous sitting, television viewing generally takes place in the home, as a solitary or familial activity, and allows for various distractions such as channel hopping and other domestic activities (Connor, 1989; Allen, 1993; Morley, 1995). As cinema going constitutes an evening out, it also involves an element of anticipation and excitement, unlike television viewing which, for the majority of people, is a habitual and commonplace activity (Morley, 1995). As a result, television viewing allows for a more casual and distracted degree of attention.[22] In fact, Morley (1995) claims that many people switch on the television without necessarily intending to watch it; it is simply a part of the domestic environment and, in some ways, this bears more similarities to the aural context of radio than the visual element of cinema. This is largely because the predominant function of television is to pass on verbal information. Whether it is in the form of fictional narratives, news programmes, documentaries or light entertainment, the spectator is able to receive and follow this information without necessarily having to watch the images. This is reflected in the way in which people are purportedly able to 'watch' television while engaging in other activities (Connor, 1989; Morley, 1995). Consequently, this has interesting implications for video dance.

It is suggested earlier in the chapter that television is dominated by realist practices and narrative texts and, accordingly, the design element is generally dependent on the context of the narrative. Even with non-fiction genres, such as news programmes or quiz shows, the choice of design is determined by the codes and conventions of those genres. To some extent video dance has a far more open brief. As it is primarily a visual genre, its immediate aim is to seize and maintain the spectator's eye. Like all television, it is somewhat limited by the relatively poor-quality image of the television medium (Monaco, 1981) and programme makers must constantly be aware that the image may be relayed through anything from a 12-inch, black-and-white portable through to a large-screen, 'high-definition' television set. It is a limitation that video dance needs to overcome. Unlike other television genres that can rely on the transmission of information via aural signification, video dance must capture the spectator's visual attention (Lockyer, 1993). In some ways this is similar to television advertising, which aims to 'seize' the viewer's eye in order to 'show' and demonstrate its product (Evans, 1988; Goldman, 1992).[23] This section therefore considers some of the strategies that video dance uses to arrest the spectator's visual attention and examines how this practice is situated in relation to postmodern stage dance and established television conventions.

A fundamental aspect of design in video dance is the choice and use of colour. Evans (1988) argues that, as television predominantly employs colour images, the use of black and white in television commercials is distinctive. Many video dance pieces are also filmed in black and white and this creates various effects. For instance, *Monologue* is the aforementioned black-and-white film of the 'talking head'. It is not the sleek and attractively presented female facial shot that generally characterizes television, but one that is homely and aggressive. The use of black-and-white filming reiterates this break with convention, as 'talking heads' are typically situated in the bright, synthetic colours of contemporary television. Another example is *L'Envol de Lilith*, which is styled as a black-and-white silent film. In this instance, in addition to the 'crime of passion' subject matter, the Western representation of orientalism, the dramatic facial expressions and the use of inter-titles to set the scene, the black-and-white filming technique alludes to the codes and conventions of the cinematic silent era. As well as these two video dance works, several others are also shot in black and white, perhaps to give the works a more cinematic or 'art house' feel, so that the video dance genre has closer associations with a high art aesthetic than with the commercial status that characterizes the television medium.

The use of colour images, however, can equally contribute to creating a strong sense of visual design. For example, in *Le Spectre de la Rose* the male performers wear suits made of various shades of pink rose petals. For some of the time the performers are filmed against a blue summer sky and white fluffy clouds. The delicate pastel hues are set in subtle contrast with one another. At other points, the dancers lie across a bed of autumn leaves. The pale pinks of the suits merge with the darker reds of the leaves, leaving the screen awash with a motley blanket of monochromatic colour. The visual effects are stunning. Another example is *Outside In*, which makes striking use of earth tones set in contrast with lush green fields and a sky-blue backdrop. The performers wear loose tunics and trousers in shades of rust, black and chocolate brown. In some instances this is complemented by a sand-coloured 'fingerprint' across which they walk, or co-ordinated with the verdant landscape against which they dance.

Another feature of design is in relation to costume, and a striking use of dress is one of the ways in which several video dance makers have succeeded in drawing attention to the visual component of their work. *KOK* is based on the sport of boxing and this idea is carried through into the costumes. The dancers' outfits are not simply replicas of typical boxing wear, however, but an elaborate and stylised parody of generic 'strong man' imagery: leopard skin shorts, wrestling vests, studded belts, black eyes, tattoos, and brightly coloured, balloon-sized boxing gloves. It is also of interest that fashion designer Jean-Paul Gaultier designed the costumes since this gives the work both a degree of high art or *haute couture* kudos and reiterates the importance of visual design. Choreographer Lea Anderson, whose work in video dance includes not only *Le Spectre de la Rose* but also *Waiting, Joan*, and *Perfect Moment*, is noted for her attention to visual detail (Briginshaw, 1995–96; Dodds, 1995–96). The costume designer with whom she regularly collaborates is Sandy Powell, who has designed for a number of other genres in which visual image plays a fundamental role: these include the work of performance artist Lindsay Kemp and several feature films, such as Sally Potter's *Orlando* (1992) and Derek Jarman's *Edward II* (1991). During one section of Anderson's *Perfect Moment* the female performers are dressed in magnificent silk ball gowns. Each one is in a different hue, and the range includes a deep orange, a dark purple and a rich cerise. Such bold colours create a remarkable contrast against the brilliant white of the studio setting and, as Briginshaw (1995–96) suggests, there are notable allusions to the visual aesthetics of television advertising.

In terms of set or location (as examined in Chapter 3) numerous video dance makers have taken their work outside the predictability of the dance studio and into a number of alternative venues or locations. The placing of a dancing body in unexpected surroundings is perhaps another way in which video dance seizes the spectator's eye. For instance, *Keshioko* takes place in a dockyard, *Le P'tit Bal Perdu* is located in a field of tall green grass, and *3rd Movement* is set in a deserted chapel. These settings are sparse so that a clutter of background objects does not distract from the moving body. In many ways this is similar to stage dance in which set designs tend to be minimal. Not only is the video dance body often situated in unexpected locations, but also, in several cases, it is presented within circumstances that are illogical and peculiar. This again is another way in which the image can arrest the spectator's visual attention. In the final section of *Codex* a man stands in darkness. As he attempts to move he is seen to have a large, flat square attached to the base of his shoe and to be wearing enormous rubber dungarees and a strange-looking neck brace. He brings his foot into the air and on the floor, where his foot was previously, is a yellow, neon square, identical in size to the one on his foot. He takes several more steps in a rather awkward fashion because of the appendage on his left foot. Suddenly, a grid consisting of nine squares divided by thick black lines lights up on the floor. There are three red squares, three yellow and three white, and the rest of his surroundings remains in darkness. Each square is the size of the square attached to his foot and as he steps on to each colour, or across several colours, the square on his foot is lit up with these same colours. There is no logical explanation as to the purpose of these events, though it is comical to see the performer battle with the peculiar appendage. Again, the visual image is stunning.

A similar series of bizarre events can be seen in *Le P'tit Bal Perdu*, *Outside In* and *Cover-up*. The way in which these images are captured on screen creates an innovative 'televisual canvas'. In *Le P'tit Bal Perdu* a male and female performer sit, face on to the camera, at a table in a field of grass. Bottles of milk and telephones appear from nowhere, oranges drop out of the sky and shovels are produced from under the table. Likewise, in *Outside In*, one of the dancers peels back a piece of turf to reveal the faces of three other dancers framed by the rectangular screen. In *Cover-up*, a 'snow storm' of alphabet letters takes place and the performer falls into a 'milk bath' in the shape of a human body. In each of these examples it is the peculiarity and unpredictability of the image that can potentially attract the spectator's gaze.

In view of the postmodern stage context out of which it has developed, it is perhaps not surprising that video dance demonstrates a concern for stylised costume, a rich use of colour, strong visual backdrops and arresting images. The employment of visual spectacle, stylized images and ornamentation are characterized as typical features of postmodern stage dance (Banes, 1987; Mackrell, 1991). It is also significant that several video dance practitioners have referred to other areas of distinctive visual imagery that have influenced their work: David Hinton draws attention to the physicality of Kung Fu movies; Alison Murray cites the popular images of cartoons and music videos; and Margaret Williams refers to her visual art background. It is suggested earlier that, although television is a visual medium, it has more similarities to the aural context of radio. As dance is clearly a visual medium, it could be suggested that video dance is drawing attention to the visual properties of television. The focus on the visual is also in keeping with various postmodern conceptualizations of television that highlight the emphasis on aesthetic spectacle and surface image that typify postmodern television texts (Harris, 1996). With video dance, the television screen becomes a site of artistic interrogation.

From the perspective of a classic television framework and through its disruption of realist practices, narrative structures and psychologically motivated characters, video dance clearly challenges accepted codes and conventions. Video dance practice, with its emphasis on aesthetic spectacle and free play of signifiers, is far closer to postmodernist characterizations of television. Indeed, it is perhaps not surprising that video dance bears the hallmarks of postmodernism in that it emerges from a postmodern stage dance context. Yet this is not to suggest that video dance should simply be conflated with other postmodern television texts. Harris (1996) differentiates a notion of 'commercial television' from an artistic or experimental avant-garde. He argues that television texts such as *Miami Vice* and music video, which promote consumerist practices, can be located within a commercial postmodernism whereas experimental art video fulfils a solely aesthetic function. 'It is important to defend other kinds of artistic experiment that genuinely aim at pushing back artistic boundaries purely for aesthetic purposes' (p. 171).

It must be noted, however, that the discussion of the avant-garde in relation to postmodernism is something of a theoretical paradox and this tension is highlighted in the problematic task of locating the underlying aesthetic of video dance. The practice of disrupting boundaries and breaking away from existing traditions, as is characteristic of video dance, is typical of the avant-garde. Yet whereas the avant-garde is a modernist concept, video dance stems from a postmodern stage dance tradition. Postmodernism is less about breaking away from existing traditions than effacing boundaries and appropriating eclectic and fragmented components from elsewhere. It could, however, be argued that although video dance clearly disrupts artistic boundaries and is experimental, or even avant-garde, in nature, this is the result of a postmodern lineage rather than any underlying modernist aesthetic. As it exists at the creative interface between dance and television, it would appear that video dance both appropriates and abandons components of dance and television traditions, a strategy that is far closer to a postmodernist aesthetic.

The findings of Chapters 3 and 4 would suggest that video dance is situated in a distinctive position. First, it transgresses the concept of dance as defined by live performance: the televisual apparatus constructs dancing bodies that transcend the capabilities of the stage body; video dance employs movement styles that are enhanced by, and compatible with, the television context; and it facilitates spectator positions that could not be achieved on stage. Second, video dance resists and disrupts many of the codes and conventions of established television texts: it challenges realist practices, narrative structures and psychologically motivated characters in order to achieve striking and unpredictable images. Consequently, it could be argued that this experimental form has called into question the way in which dance and television are each conceptualized and engaged with. Critics who dismiss video dance for its lack of established dance techniques need to reconsider the evaluative criteria of what constitutes 'dance'. It is perhaps no longer appropriate to analyse video dance within stage dance criteria; a whole new set of appraisal techniques needs to be developed to take into account the role of the televisual apparatus in the determination and construction of movement. Similarly, choreographic practices have had to be reconceived to take the televisual apparatus into account. It could also be suggested that video dance requires the spectator to view television in new and alternative ways; video dance has the potential to draw attention to the visual, aesthetic and formal properties of television, which have tended to be overlooked or marginalized within mainstream television. Video dance constitutes a type of 'art television' through which aesthetic innovation is explored. It is the hybrid status and experimental agenda of video dance that allows it to challenge and disrupt established boundaries and, as a consequence, problematize existing aesthetic and conceptual structures.

5
Hybrid Sites and Fluid Bodies

Video dance and hybridity

Video dance is a creative exploration of the relationship between dance and television. It is a fusion or amalgamation of two distinct sites in which the codes and conventions of each medium are inextricably linked. Yet this duality often fails to be taken into account. It is suggested in Chapter 1 that a particular body of critical writing has either dismissed or overlooked the televisual aspect of video dance and thus completely disregards an intrinsic component. Meanwhile, it is noted in Chapter 2 that, with certain genres of screen dance, the interception of the film and televisual apparatus is underplayed or kept to a minimum, or that creative innovation has given way to commercial appeal. Thus, in recognition of the merging of dance and televisual practices in video dance, the concept of 'hybridity' is a useful characterization. The proposition that video dance constitutes a hybrid site both acknowledges and reiterates the innovative interrelationship between dance and television.

It is significant, however, that the fusion of dance and televisual practices in video dance is far from straightforward. The findings of Chapters 3 and 4 highlight a dialectic contest between the codes and conventions of each medium through which a creative tension occurs and, consequently, boundaries are challenged and displaced. Televisual devices act on dance in such a way that bodies are constructed that transcend the capabilities of the live dancing body and that can only exist within the television context: in turn, formal strategies from the postmodern stage dance tradition, out of which video dance has emerged, act upon the television medium to resist and subvert established realist practices and to create striking aesthetic images not normally associated with conventional television texts. This hybridization clearly has an impact on choreographic practices, spectatorship positions, the role of the performer and the television medium as a context for dance. Thus video dance is marked by a disruption of symbolic boundaries.

It appears that the blurring and displacement of boundaries in video dance is not solely in relation to dance and televisual practices. It could be argued that video dance slips into other aesthetic sites and theoretical frameworks. Therefore the focus of Chapter 5 is to consider two areas that are particularly pertinent to the interdisciplinary character of video dance. The first is the consumer imagery of television advertising and music video and its network of promotion, and the second is in relation to discourses of technology. The way in which the video dance body traverses these different material sites and theoretical disciplines suggests a certain fluidity. Hence the concept of a 'fluid body' is a significant theme for Chapter 5.

Video dance, television advertising and music video

A number of dance critics have drawn comparisons between video dance and both the commercial advertisements that are screened on television and the promotional videos that accompany many popular music singles (Bayston, 1992; Maletic, 1987–88; Bozzini, 1991; Meisner, 1993; Jordan, 1992).[1] Yet there has been no sustained or in-depth analysis that rigorously seeks to address this comparison and any examination that does take place is somewhat superficially dealt with in the space of a few lines. For instance, Meisner (1991) refers to the 'upbeat, contemporary presentation' of video dance, which 'should surely appeal to our trendy youth, fed on pop videos' (p. 17) and Bayston (1987) notes, in his analysis of the *Dancelines* project, that 'many of the technical tricks are used in commercials and pop promotional videos' (p. 707). Similarly, Rosiny (1994) states that video dance 'paralleled the MTV boom from the middle to the end of the eighties' with 'the video-specific possibilities of electronic image modification, fast cutting and effects' (p. 82).

It is perhaps not surprising that there are apparent similarities among the three forms, in that video dance and music video largely evolved during the 1980s, and television advertising, although it has a considerably longer history, underwent several major changes in the late 1980s during which it became less 'routine' and far more ambiguous (Goldman, 1992). All three are characterized by eye-catching visual images and the body is central to each form: in video dance the body is a primary component; in television advertising bodies are used to promote and display the related commodities; and in music video the bodies of singers, musicians and dancers pervade the genre. What is pertinent to this study of video dance is that the bodies are positioned, presented, moved and filmed in innovative and stylized ways. Significantly, video dance, television advertising, and music video have all been conceptualized within a postmodern framework. Video dance has emerged from a postmodern stage dance tradition and certain television advertisements and music video texts have been characterized as typically postmodern (Wyver, 1986; Kellner, 1995; Harris, 1996).[2] The postmodern context, with its emphasis on visual spectacle, is clearly a point of commonality (Jameson, 1991; Goldman, 1992). Therefore the aim of this section is to address areas of similarity and overlap between video dance, television advertisements and music video.

In Chapter 2 it is suggested that advertisements are structured in such a way that the commodity is linked to a particular image or association. Those advertisements which employ explicit images of dance, such as the Cadbury's Twirl and Hellmann's mayonnaise examples, tend to employ popular stereotypes and conventional dance techniques. It would therefore appear that video dance has little in common with these commercial representations of dance. Yet there are other advertisements that do not contain any established dance vocabularies but which share certain similarities with the way that the body is constructed, manipulated and distorted in video dance.

One example of this can be seen in an advertisement for Ariston, the brand name of an electrical, household goods manufacturer. The action is located in a type of avant-garde apartment which combines the look of a continental villa with a richly coloured, minimalist loft. The apartment consists of a round window, a balcony, a staircase and a free-standing, 'arty-looking' goldfish bowl. The advertisement features a washing machine, a fridge-freezer, and a cooker and is set to a repetitive, synthesized melody. The copy that appears across the base of the screen, and is repeated by a monotonous, male voice-over, states, 'Ariston, and on and on and on ...', until the end of the advertisement when it draws to a close with, '... and so on'. The visuals are based on an accumulation technique in which various 'characters' repeatedly come into the space and, each time, perform identical actions to their previous appearance. For instance, the piece begins with a close-up of the washing machine. The camera zooms back as a young woman rushes up to it, removes a sparkly pink skirt from inside the machine, puts it on, and dances off-screen with a young man who has just skipped towards her. Seconds later, this precise image is repeated, but now also includes a young boy who hobbles across the screen with his leg in plaster, supported by a crutch. More and more characters gradually enter into this surreal landscape: a window cleaner keeps falling from his ladder, a cycle courier puts a letter in the oven, a person dressed as a fluffy rabbit bounces over to the fridge to retrieve some carrots, a large woman balances a wobbly pink desert on a plate, and a dog repeatedly scampers through the chaos. The advertisement ends with the image 'freeze framed'. There is no suggestion of any type of resolution; instead, the final words 'and so on' suggest that the action could continue.

To some extent, this formalist device, which accumulates bodies within a repetitive cycle of movement, is extremely similar to the video dance piece *Tango*, in which numerous people enter and exit a room (as described in Chapter 3). There are further comparisons in that both are located in domestic settings, both draw on pedestrian activity, and the 'performers' consist of 'everyday people' dressed in 'everyday clothes'. Yet, whereas the accumulation device in *Tango* is purely a play of formalist apparatus, the use of cyclical action in the Ariston commercial is closely tied to the commodity. First, the connection between Ariston household durables and repetitive cycles suggests that these goods can be reliably used over and over again. Second, because the 'visual cycles' of the advertisement are constructed through 'special effects technology' and draw upon arty, surrealist imagery, Ariston products are given a high-tech, avant-garde status. The Ariston name, and the household goods that it represents, thus become associated with 'state of the art' technology. So although there are clear similarities between the design and movement of the body in both the Ariston and *Tango* video, the former is closely tied to a commodity-sign whereas the latter is not.

A similar example arises with an advertisement for the Nationwide Building Society. The action takes place in a white studio setting, although it is designed like a bank counter, with a large blue Nationwide sign in the background. The images are set to a comical, rhyming song that lists a variety of people from different places. For instance, the first verse states, 'Cooks in Cookham, Bakers in Burnham, Don't leave the loaves long, Don't forget to turn 'em'. The visuals meanwhile consist of pixillated[3] images that illustrate the lyrics. Like the Ariston commercial, the result is somewhat surreal: bakers carry their loaves of bread, a tractor appears with a ram on the back, two men carry a bath tub, a potter sits at his wheel and, at one moment, twenty or so dogs race across the screen. Not only are the activities incongruous with the 'bank setting', but the pixillation effect makes the bodies appear all the more bizarre and comical as they zoom around the frame. The advertisement ends with a cut to the Nationwide sign against a plain white background as a male voice-over states, 'Nationwide, with more branches than anyone else, we're the nation's building society'.

In terms of the way in which the body is constructed, this advertisement has similarities to video dance. For instance, the pixillation technique is also used in *Topic II/46 BIS*, in which the dancers appear to 'skate' through the streets of Prague (as described in Chapter 3). Within both the advertisement and the video dance piece, the body is able to move in ways that are physically and temporally impossible for the 'live body' to achieve. It is established in the previous two chapters that the video dance body has a tendency for the unpredictable: its movement capacity is more versatile as a result of special effects; the editing can allow for spatial, temporal and dynamic possibilities that could not occur outside the television context; and there is a discernible resistance to realist strategies, which results in striking and unusual images. It could be argued that the images from the Nationwide advertisement pertain to these characteristics. Yet although on the surface these bodies appear strange and unpredictable, the images closely follow the lyrics of the song. Within the logic of this advertisement, the actual bodies are placed in a coherent framework of meaning.

Whereas *Topic II/46 BIS* solely manipulates formal televisual properties, the Nationwide commercial is closely tied to a commodity and the interpretive logic of advertising. The promotional message of this particular advertisement operates on various levels. First, the relationship between the lyrics and the image simply reinforce the notion that Nationwide attracts a diversity of people from wide-ranging professions and geographical locations. Second, the static Nationwide sign, which is ever present while the 'characters' whizz back and forth in chaos, implies that the building society remains stable and permanent. Finally, it could also be suggested that the bizarre and comical bodies provide a memorable image for the consumer. Cook (1992) describes this type of mnemonic strategy as a 'tickle ad' in which a commodity becomes memorable through distinctive or humorous promotion. Although the advertisement may explore certain formal devices that are comparable to video dance, these strategies are closely tied to the product.

Another advertisement that deserves some attention in relation to video dance is one of the Guinness campaigns, and it operates as both a 'tickle ad', due to the comical action, and as an 'ambiguous ad', in that one level of meaning is dependent on readers who are 'in the know'. It is set to a piece of slapstick music and features a bartender and a young male. The action switches between the bar counter and what is presumably the bar area, although it is not of a realist décor. The design is stark and the colour consists of pale greys and blues, black and gold. The sales pitch of the advertisement is based on the belief that, in order to pour a decent pint of Guinness, the drink has to settle before it can be filled up to the top of the glass and, even then, it has to settle once again. The advertisement accordingly plays on the anticipation of the consumer as he desperately waits for his much-desired pint. The young man runs, jumps, *bourrées* like a ballet dancer, swings his hips, twiddles his fingers, cracks his knuckles and springs towards and away from the glass of Guinness. Rapid changes of shot neatly sum up his sense of frustration as he waits for the drink to settle. There is also a regular focus on the drink itself, which is highlighted through vertiginous shifts in perspective. The camera position switches between different shot sizes and distances so that at one moment the pint of Guinness takes up most of the screen and is equal in height to the consumer, while at others it is small enough to be hand held. There is a strong sense of visual design, alongside an unpredictability of how the image will be framed, where the body will appear next and what it will do that makes it similar to much video dance.

Once again, however, the various components of the image persistently refer back to the commodity-sign. The Guinness name is written across the front of the pint glass, which is continually in view. The comical and bizarre movement behaviour of the young man creates an image, which first, serves as a key reminder of the Guinness brand name and, second, differentiates the product from other alcoholic beverages that are perhaps promoted through more conventional realist strategies. In fact, Guinness has a tradition of innovative advertising. The 'Pure Genius' trademark, accompanying all Guinness advertisements, refers not only to the 'excellence' of the product but also to the memorable creativity of previous campaigns. Yet although the young man's actions could be considered zany, they have a coherent logic within the context of the advertisement. While this reading can only be accessed by those who already consume Guinness and know that the drink needs to 'rest' for several moments, the protagonist's actions are based on frustration and anticipation as he waits for the drink to settle. Likewise, the use of colour is not simply arbitrary but refers back to the black-and-gold Guinness trademark and to some of the greys, blues and golds that are picked up on camera as the creamy pint is poured.

In terms of the use of colour and filming devices, a similar issue occurs with an advertisement for Gordon's Gin. The image depicts rapidly edited fragments of a male body as it flies down a perspex chute into a pool of clear liquid and is then surrounded by an explosion of green-and-white fizzing bubbles. Another feature of video dance (addressed in Chapters 3 and 4) is the use of fragmented close-ups and eye-catching visuals. In this advertisement, however, the colour scheme refers back to the trademark colours associated with a 'Gordons and tonic', the body that flies 'down the chute' is analogous with the act of drinking/consuming, and its immersion into the clear, sparkling liquid conjures up the refreshing and invigorating experience of consuming a Gordons and tonic.

The above analyses would suggest that there are key similarities between certain television advertisements and video dance texts. This is less to do with a discernible dance element than with specific formal characteristics, particularly in relation to the treatment of the body. For instance, the body is often situated in striking visual images, surrounded by rich and glossy colours, and positioned and filmed in innovative ways. The movement images are largely constructed through the televisual apparatus and could not exist outside this context. Thus it could be suggested that the video dance body shares, and is inscribed with, some of the codes and conventions of the body in television advertising. Yet there is one fundamental difference between video dance and the advertising form.

Advertising is used to promote a particular commodity, whereas video dance is not. Yet this differentiation is brought into question with the *Paramount Hotel Ads* (choreographer/director: Pascale Baes, 1994). Although they have been screened during 'commercial breaks' on US television, they have only been shown in the UK as part of the *Tights, Camera, Action! 2* series. The decision to place these advertisements in a video dance series highlights the aesthetic similarities between video dance and television advertising and calls into question the role of the body in these two forms; for instance, is the advertising body intended purely to fulfil a promotional function, and the video dance body an aesthetic one, or is this too simplistic a division? Just because the video dance body is not overtly linked to a specific product, this does not necessarily imply that it lacks any promotional or commercial potential. *The Paramount Hotel Ads*, and their ambiguous promotional/artistic status, are addressed later in the chapter, but for the moment attention turns to music video and its relationship to video dance.

In Chapter 2, the music videos *True Faith* and *She Drives Me Crazy* by choreographer and director Philippe Decouflé are the subject of investigation. As a comparison to Decouflé's work in music video, it would now be useful to consider his work in video dance in order to explore any overlap between the two forms. Therefore I intend to look at two pieces, both of which are referred to in Chapters 3 and 4. The first, *Le P'tit Bal Perdu* features a man and woman who sit behind a table in the middle of a field of long green grass. Occasionally there are cutaway shots of some black-and-white cows and of a stout elderly woman who sits playing an accordion in a field surrounded by a string of white lights. The action is set to a French folk song, 'Le P'tit Bal Perdu', which can be translated as 'the little lost ballroom (or dance)', and the lyrics describe a ballroom, the name of which the singer cannot remember. The use of pun is a central device throughout the work.

In Chapter 2 Goodwin's (1993) theory of the 'musicology of the image' is employed in the analysis of music video. The same concept lends itself well to *Le P'tit Bal Perdu* in that both the form and the content of the movement closely follow the structure and lyrics of the song. The choreographic content is made up of a complex, gestural motif (similar to the 'rubber shell' woman in *True Faith* with her elaborate sign language) and the choreographic form mirrors the structure of the song; identical movement motifs are repeated when the chorus (or phrases in the verse) is also repeated. The actual movement closely corresponds with the lyrics to the song. In some instances, this is through mimetic representation. For example, on the word *guerre* (war) the performers point their fingers like a gun and on *deux* they hold up 'two' fingers. At other times the gesture is more abstract; yet it is still possible to read the links between word and image. For instance, with the word *importante* the performers stand up with their hands on hips as if to demonstrate authority, and on *piste* they mark out a winding trail with their hands as if to represent a track. A number of incongruous props also feature in the piece and tie in with the punning device. The words *qui s'appelait*, meaning 'whose name was', are repeated several times throughout the song, and each time a pun is brought into play around the '...pele' or '...ler' sound. A telephone appears on the table as a reference to *appeler* (which can mean 'to call'), along with several bottles of 'milk' which are a play on the *lait* sound. This occurs again with *pelle* as a variety of shovels appear, and *laid*, at which the man pulls an ugly face. The work is a play on the problematics of semantics.

This piece displays several similarities with Decouflé's work in music video. The strong sense of visual design, the gestural vocabulary, the use of humour, the absurd tasks and the incongruous images are all recognizable hallmarks of Decouflé's style. Yet there is one fundamental difference in that, although the image closely follows the music, there is no promotional strategy connected to the song. The viewer is told that the music is a traditional French folk song, but it is not credited to a songwriter or performer. Although there is an internal logic to the relationship between words and image in *Le P'tit Bal Perdu*, there is arguably no external agenda. In *True Faith* and *She Drives Me Crazy*, the images both visualize the song and, as with advertising, promote various messages and associations in relation to the band; in *Le P'tit Bal Perdu* the images are simply a postmodern play of signifiers.

This is also very evident in *Codex*. The piece is made up of three sections that have no obvious connection except that a set of hieroglyphics, accompanied by an incomprehensible male voice-over (which is possibly played backwards) precedes each section. The first section features a group of dancers covered from head to toe in black wet suits and wearing flippers. The second section consists of a sensitive contact duet between two men, accompanied by a haunting melody, sung by a nearby figure dressed in enormous rubber trousers and a metal neck brace. At one point he is drenched in a shower of water and later he is shown suspended upside down from the ceiling. The final section features the man in the neck brace again. His trousers are full of water and he has a large square attached to his foot. Close by, a grid lights up that consists of brightly lit red, white and yellow squares. Depending on where the man treads, the square on his foot lights up to match the square or squares that it covers on the grid.

Many of the stylistic hallmarks of Decouflé's work in music video are also apparent in *Codex*. For instance, the employment of curious costumes, such as the flippers and the rubber trousers filled with water, which hinder or affect the performer's movement, also occurs in his music videos. There are the three dancers in *True Faith* who wear the bloated, Oskar Schlemmer style outfits which limit the articulation of their limbs. The use of absurd costumes and seemingly fruitless tasks, such as the man with the square attached to his foot in *Codex*, creates a point of humour in terms of the struggle against impossible odds. Likewise, in *True Faith* there is the man who tries unsuccessfully to stack the three geometric shapes. The way that the televisual apparatus manipulates or distorts movement in video dance is also typical of Decouflé's work in music video. In the first section of *Codex*, at one point the image is rotated 180 degrees so that the bodies appear to be suspended in the middle of the television screen, and later the image is speeded up causing the bodies to whizz around at a physically impossible rate. Other devices confuse size and perspective. In one instance, the camera is placed in close proximity to the man in the rubber dungarees, who sings the melody and yet is some distance from the men's duet. As a result, the singing man's cheek and mouth take up the whole of one side of the screen, while the other two men can be seen in full body shot on the other side of the frame. A similar manipulation of camera work and editing occurs in Decouflé's music videos: there is the rapidly edited images of Roland Gift as he appears to zoom towards and away from the screen in *She Drives Me Crazy* and the slow motion and speeded up antics of the three rotund characters in *True Faith*.

It is a combination of peculiar events, a rich use of colour, striking spatial design, and an innovative relationship between the moving body and the televisual apparatus that creates this series of arresting images, and it is possibly for this reason that Decouflé's work is not only compatible with music video but is also said to be be stylistically similar to advertising (Bozzini, 1991). Yet, as with *Le P'tit Bal Perdu*, *Codex* is not bound to a commodity. It is not restricted to a promotional function; whereas *True Faith* and *She Drives Me Crazy* are rooted within popular music conventions and conform to an element of closure, *Le P'tit Bal Perdu* and *Codex* are able to resist this element of coherence and logic. Textual radicality can be an end in itself. Yet there seems to be a danger of oversimplification in the assumption that, as video dance is not directly attached to a discernible commodity, it resists the promotional discourses that characterize television advertising and music video; there is a supposition too that just because music video and television advertising are linked to a commodity, their function can only ever be promotional. The following section addresses these issues through an examination of the promotional context of music video, television advertising and video dance, and a discussion of the extent to which the body in video dance can be posited as a 'consumer body'.

The consumer body and the promotional network

The stylistic similarities among video dance, music video and television advertising are clearly evident in their construction of striking visual images, in which the body is a central component. Dyer (1988) describes the way in which advertising employs technologically sophisticated framing devices, lighting, lenses and colour to create eye-catching imagery, and this can also be seen in video dance and music video. As the television screen depicts a relatively poor-quality image, these three forms are required to find innovative visual formats that attract the viewer's gaze. Kaplan (1988b) notes that music video borrows visual conventions from advertising and O'Donohoe (1997) draws attention to advertising's appropriation of ideas and techniques from other art forms, such as cinema, pop video, comic strips, television programmes and fine art. These ideas are further supported by Goodwin (1993) who states that the emergence of music video coincided, both textually and institutionally, with the convergence of various media sites in the 1980s. Instances of the crossover of media texts are seen in pop music used in advertising and star names used to endorse products, or cross-media marketing deals that might incorporate the rights to a film along with a soundtrack and related merchandise. O'Donohoe (1997) conceptualizes this intertextuality in terms of 'leaky boundaries' to describe the way in which popular texts merge and overlap. Indeed, it is perhaps not surprising that certain formal similarities emerge in that there is a crossover of personnel within these areas: filmmakers have entered the commercial world of advertising and music video (O'Donohoe, 1997; Nava, 1997) and various dance practitioners, such as Alison Murray, Lea Anderson, Philippe Decouflé and Wayne MacGregor,[4] have been creatively involved in music video and television advertising.

Yet irrespective of these stylistic similarities among video dance, music video and television advertising, there is one fundamental difference: the bodies of advertising and music video are closely tied to a 'commodity' whereas the video dance body is not. The body in music video is used to refer to and promote a record, CD or cassette and the advertising body is inscribed with discourses that are transferred onto a product. This would suggest that television advertising and music video are concerned with the construction of a 'consumer body'; a body that is commodified and promoted. Yet, although the body in video dance is not tied to a discernible commodity, with a given 'exchange value', it could nevertheless still be conceptualized as a consumer body. After all, video dance undoubtedly exposes and promotes the work of individual directors, choreographers and itself as a genre. As with any artistic practice, to some extent, it fulfils a promotional role.

Although television advertising and music video texts are concerned with the promotion of a discrete commodity, it has been suggested that 'advertising' is a commodity *per se* that functions to promote itself as a social form and to promote consumerism as a social practice (Wernick,

1991; Goldman, 1992). It is also apparent that television advertise-ments are not isolated texts. Not only do campaigns for a certain prod-uct recur across several media, but advertising in general draws on shared values and promotional conventions (Williamson, 1978; Goldman, 1992). Indeed, Wernick (1991) proposes that the intertext of promotion has gone radically beyond the advertising media into all aspects of cultural life. He states that promotion is a signifying system applicable to commercial and non-commercial phenomena: institu-tions such as religion, politics and education employ forms of promo-tional discourse; cultural goods are situated within a vast promotional web of commodified lifestyles; and the individual is unavoidably sub-ject to self-promotion. Wernick (1991) goes on to note that as areas of social life have become increasingly commodified, promotion is now a 'cultural dominant'.

Wernick (1991) presents various strategies through which a 'com-modity' may promote itself and many of these are applicable to video dance. First, he introduces the notion of 'hooking', an eye-seizing strat-egy designed to be a self-advertisement. One component of video dance, identified in Chapter 4, is its 'need' to capture visual attention through such devices as a rich use of colour, a striking design element and arresting images. These devices are perhaps 'hooking' strategies in that video dance could be said to promote itself through placing the body within distinctive and unconventional imagery. Another mode of promotion is through the serial form. The aim is to build up a market through serializing a commodity and this is applicable to both the *Tights, Camera, Action!* and *Dance for the Camera* programmes. By the very nature of the serial form, each episode promotes the following episode; yet each is self-contained. There is also the notion of 'promo-tion through name'. Although video dance does not employ 'star names' as such, the use of certain choreographers in video dance can act as a promotion for their stage work.

Wernick (1991) refers to 'promotional transfer across commodities' to describe the way that commodities can in turn promote other com-modities. Although it appears to be on a relatively small scale, there are instances when video dance can demonstrate this idea. For instance, *3rd Movement* and *Monologue* respectively employ the music of Tchaikovsky and Monteverdi, which not only promotes the individual composers but also classical music *per se*, and the use of designer Jean-Paul Gaultier in *KOK* draws attention to both his fashion label and to 'haute couture' in general. Wernick (1991) concludes with the notion of 'promotion without end' to describe the way in which promotion consists of endlessly proliferating signifiers: 'Each promotional message refers to us a commodity which is itself the site of another promotion' (p. 121). The concept of 'promotion without end' would therefore suggest that irrespective of whether a cultural sign or commodity is 'framed' as a promotion, it has the potential not only to promote itself but also to refer to a whole network of further promotional signs. If this is the case, then it appears that the video dance body is unavoidably inscribed in promotional discourses.

The significance of the body in late-capitalist society is an essential feature of consumption (Featherstone, 1991). The body provides the basis for a vast array of body-related products, which are promoted by both the 'official bodies' of advertising and 'everyday bodies' that are inscribed by these various corporeal regimes. In advertising there is a primacy of surface display as demonstrated through the idealization of health, youth, beauty, and slimness (Featherstone, 1991). Consumer culture thus provides constant images of the body as a means to stimulate and maintain consumer desire. Grossberg (1993) suggests also that the body plays a major role in the visualization of music in terms of clothes, make-up, hair styles, sexuality and dance. Pertinent to this dominance of the 'body beautiful' is that one of the characteristics of advertising (and music video) is its fragmentation and commodification of various parts of the body. Goldman (1992) suggests that advertising has abstracted various areas, such as lips, eyes, and hands, which carry their own set of conventionalized discourses. This formal device bears similarities to video dance, which also fragments areas of the body through the use of close-up camera work. There is one major difference, however, in that the fragmented advertising body carries a set of messages that are then attributed to a commodity, whereas the use of fragmentation in video dance is simply a play of the formal apparatus. In some instances this may be a televisual convention, such as the use of facial close-ups to depict expression or emotion, but it is never tied to a commodity other than the video dance body itself.

Although there are considerable similarities between the conventional stage dance body and the consumer body in the sense that both are youthful, healthy, slim and attractive, this comparison is inappropriate to the video dance body. In video dance the element of 'dance' occurs as much through camera work, editing techniques and special effects as through the actual body, and allows for bodies of varying size, shape, weight, age and ability. The video dance body could be said to have a 'promotional agenda' in so far as it promotes various conceptualizations of a 'video dance body', but these constructions are not necessarily restricted to the limited boundaries of the consumer bodies in advertising that are always youthful, slim, beautiful and so on. For instance, *Outside In* employs bodies that are physically disabled (or rather 'differently abled'), *Monologue* depicts a woman's face that shouts aggressively and sprays drool towards the camera, and in *Kissy Suzuki Suck* the two women direct obscene sexual gestures to the spectator and defile their arms and faces with lipstick 'graffiti'. Notably, many of the bodies in video dance are 'absurd bodies'. They are situated in zany imagery and move in ways that are impossible for the live body to achieve. This is partly due to the 'art status' of the video dance body. Whereas it is permissible for the video dance body to be inscribed in ambiguity, the advertising body always refers back to a commodity. This implies a certain element of closure and coherence which is less applicable to the video dance body.

Throughout the 1980s, however, a style of advertising emerged that was considerably more ambiguous and open-ended than previous advertising and it is perhaps here that video dance and advertising most notably converge. Prior to the 1980s advertising had been explicit with its links between a commodity and its associated image, but during the 1980s a mode of advertising developed, characterized by its ambiguity, irony and self-reflexivity (Goldman, 1992; Nava, 1997). This is a particularly pertinent period in that music video evolved, and drew on many of the codes and conventions of advertising, and video dance also developed during this era. The stylistic developments of advertising in the 1980s have been attributed to a crisis of sign-values and the emergence of a sceptical consumer (Goldman, 1992; Nava, 1997). Goldman (1992) suggests that, due to the pervasiveness of advertising, there has been an acceleration of sign-values as advertisers feverishly attempt to differentiate their product from others. The constant overload of imagery and the saturation of promotional discourses has resulted in the development of a sceptical consumer, insulted by the blatant positioning strategies of advertising and bored by its conventions.

In response to the sceptical viewer, a style of advertising has emerged that is 'witty, sophisticated, intertextually referenced and visually appealing' (Nava, 1997, p. 46). One of the ways in which advertising and music video have attempted to differentiate themselves is through the employment of avant-garde art strategies. Buckland (1993) and Cook (1992) note that there has always been a degree of cross-influence among the world of art, popular music and advertising texts. Indeed, the recent focus on aesthetic innovation and art values in promotional imagery is perhaps reflected in the numerous industry awards that celebrate music video and advertising and the discussions of advertising that take place on arts programmes (Frith, 1993). Goldman (1992) suggests that due to accelerated sign-values and society's sophisticated reading ability, advertisements have become increasingly abbreviated: verbal text has decreased, meanings are compressed and short-cut framing conventions are used. He states: 'the commercial avant garde [is] trying to differentiate itself from the crowd by adopting eccentric framing techniques or pursuing the logic of frame reductionism to its minimalist limits' (p. 155). Goldman (1992) refers to this particular genre of advertising as the 'this is not an ad' type. The features of these advertisements are a lack of formal boundaries, an absence of coding markers, and invisibility of the commodity itself, hence the commodity-image is ambiguous. He suggests that the strategy behind these advertisements is to flatter any viewer who is able to recognize the actual advertising campaign.

A perfect example of the 'not-ad' type may be found in the *Paramount Hotel Ads*, which were screened as part of the *Tights, Camera, Action! 2* series. Their ambiguity as advertisements or art texts is highlighted by the very merit that they were shown as part of an arts programme. Three advertisements were included, and each employs roughly the same format: a series of pixillated images are followed by an almost identical 'tag line'. The first advertisement depicts a man and woman who move through a hotel lounge area and end up seated on a staircase. At the very end, the screen goes blank and the copy flashes up in a series of three statements: 'Great Lobby...Paramount. New York...From $100 a night'. The second advertisement shows a man, carrying a large birthday cake, who takes it up to a woman in her suite. This time the copy reads, 'Great room service...Paramount. New York...From $100 a night'. The final advertisement follows a man lying down as he travels by elevator up to his room and into bed; on this occasion the written text states, 'Great beds' and so on.

These advertisements display several features that have come to characterize the commercial avant-garde. Each one is barely 30 seconds in duration, which is in keeping with the minimalist statements at the end. The action is set in the actual Paramount Hotel, which was designed by Philippe Starck and is known for its cutting edge décor. In terms of the stylized environment and the pixillated bodies that travel through the hotel in a jerky and non-realistic fashion, the images are arresting to the eye. The body is not inscribed within conventional promotional imagery, but is peculiar and unpredictable. The music is electronic and ambient, which adds to the arty feel. What makes the advertisement typical of the 'non-ad' genre is that it is unclear what the advertisement is promoting. The jolting quality of the pixillated image is difficult to follow and no single commodity comes to the fore. Only at the end is it evident that the advertisement is promoting the Paramount, but even then it is never actually stated that the Paramount is a hotel. Consequently, the connection between the commodity and the sign remains ambiguous. In order to make the link between the hotel and the arty images the reader would have to know something of the Paramount Hotel, its association with Philippe Starck, and the avant-garde art world.

Goldman (1992) states that ambiguity, elusiveness and abstract, non-representational imagery are all facets of the commercial avant-garde. Constantly proliferating signifiers and polysemy are employed to create an innovative selling point for the product. This type of differentiation is also a feature of the artistic process. Innovation is one of the motivating factors behind artistic practice and, at times, this has a commercial aim. Not only are practitioners concerned to a greater or lesser extent with stimulating audience interest through new artistic ideas, but many artists also have a commercial aim in mind; they need funding to survive and some degree of remuneration comes either directly or indirectly from a paying public. With increasingly limited subsidies available, many practitioners modify their work to suit funding body policy, which seeks to maximize audience targets. This concept, however, is extremely difficult to quantify. In the same way that those involved in the creative aspect of advertising may demonstrate aesthetic concerns and artistic integrity, video dance makers may have commercial interests and audience accessibility in mind.

There is clearly a danger of a simplistic polarization of intentionality in terms of equating commercialism with promotion and aestheticism with video dance. After all, advertisements and music video can have striking aesthetic properties and video dance can be considerably motivated by promotional concerns. A more appropriate conceptualization may derive from a notion of a 'hierarchy of discourses'. Although promotional advertisements and art texts both employ artistic and commercial considerations, promotional texts are ultimately concerned with a commercial agenda, while video dance is fundamentally concerned with an aesthetic one. The art status of video dance allows it to provoke and stimulate audiences, to challenge existing traditions and to differentiate itself from other art texts as an end in itself. In contrast, promotional texts primarily employ these devices as a means to generate capital. Yet this argument may collapse when the consumer or reader is taken into account.

The rapid developments of commodity culture and the mass media have resulted in what Goldman (1992) describes as a hegemonic state of advertising. Capitalist society implicitly assumes that everyone is a potential consumer and that each time the subject makes a purchase, she or he buys into a world of commodities (Williamson, 1978; Wernick, 1991; Goodwin, 1993). This theoretical perspective conceptualizes advertising as an ideological process in which the consumer is perpetually positioned by the discourses of promotion. Williamson (1978) refers to the Althusserian notion of the 'subject' in relation to the way in which the reader is positioned each time she or he participates in the logic of advertising; this has an ideological effect for when the individual recognizes that the advertisement is addressed to her- or himself, she or he is 'hailed' by it.

Goldman (1992) suggests that critiques of advertising tend to be directed at explicit content, such as whether the consumer is given misleading information about the product or whether the advertisement uses acceptable imagery. This focus ignores the formal logic of advertising; yet it is there that the ideological work is said to take place (Williamson, 1978; Dyer, 1988). The formal framework of advertising connects a product with a certain social value, in the form of an image, a mood, an experience or an emotion. Yet the consumer never questions the logic of this connection as this process is naturalized (Wernick, 1991). Williamson (1978) describes this as an 'objective correlative' in that the link between a product and an image takes on an objective status. Williamson (1978) notes that the actual process of connection bypasses the consumer's perception, as she or he is caught up in the hermeneutic code of the advertisement.

This 'manipulationist' model of the consumer is challenged by an opposing theoretical perspective: a 'populist' conceptualization (Slater, 1997). Although it could be argued that all forms of promotion set out to position the reader as a consumer, the hegemonic potential of advertising is not as omnipotent as it perhaps appears. For instance, a consumer may internalize the underlying ideological messages promoted by certain advertisements: yet she or he may choose not to buy the actual commodity. Or conversely, a consumer may buy a product, but not necessarily take on the ideological positioning (Goldman, 1992). Although the subject is positioned as a consumer, the effectivity of an advertisement is largely dependent on the reading position that the consumer adopts. The frameworks of advertisements contain reading rules that aim to rationalize and streamline interpretations; yet preferred readings cannot be guaranteed because of unintended ambiguities, mischosen referents and poor design (Goldman, 1992). Some viewers may be 'under-socialized' in that they do not have the appropriate lifestyle to decode the message. Other viewers may be 'over-socialized' in that they have become sceptical of advertising and purposely misread, or deconstruct, the advertisement (Goldman, 1992).

Goldman (1992) takes a particularly pessimistic view of these 'resistant readings', as he argues that they are in no way detrimental to the promotional form. If the under-socialized reader is unable to read the signs of a particular advertisement, then she or he is clearly not the type of person the advertisement is aiming at. The sceptical reader has meanwhile been dealt with through a genre of advertising that Goldman (1992) describes in terms of the 'knowing wink'. These advertisements comment on advertising itself and, whether it is through light mockery or acerbic critique, they encourage 'savviness' or cynicism on the part of the viewer. The advertising industry has simply incorporated this mood of scepticism into its framework through the introduction of ambiguous, playful or self-referential advertising.

Yet Nava (1997) posits that young readers in particular consume advertisements as they would other cultural forms, such as pop singles and magazine articles, and O'Donohoe (1997) argues that many young people readily divorce the selling message from the art form of advertising. The same practices of resistance and subversion may be employed with music video for readers may admire, comment on, and engage with certain music video texts; yet they may not necessarily like the music, nor wish to buy the single. Although advertising texts are fundamentally concerned with a commercial agenda, the consumers of these texts may prioritize the aesthetic properties of these forms. This is far closer to the way in which a reader may engage with video dance. Hence, the 'hierarchy of discourse' theory is clearly dependent on how a particular text is 'framed' and by whom. Although the producers of television advertising and music video are predominantly concerned with the promotional framework of these forms, the consumer may privilege the aesthetic framework and resist the selling message.

This model of the consumer as autonomous (Slater, 1997) clearly challenges the omnipotence of the promotional form; yet Wernick (1991) posits that advertising draws upon and reinforces dominant social values. As the agenda of advertising is to sell commodities, promotional texts draw on the most conservative values in order to embrace the widest possible sector of people.[5] Advertising feeds off the

norms and values of the consumers it addresses. This, however, is by
no means a static phenomenon. As certain social values change, so too
do the discourses of advertising, by simply reincorporating oppositional or resistant ideologies into the form.[6] This once again places
advertising within a hegemonic framework.

Notions of playfulness, irony, and constantly proliferating signifiers,
which imply a subversive potential through unstable and open-ended
meaning, have come to the fore in recent postmodern analyses of popular culture; yet several theorists have challenged the validity of these
claims. Goldman (1992) posits that the polysemy and lack of closure
that characterize the commercial avant-garde are no real challenge to
hegemonic ideology. Instead, he suggests that such texts are simply
another consumerist device intended to negate notions of conformism.
In relation to music video, Goodwin (1993) notes that postmodern
readings have tended to address textual radicalism rather than notions
of social power. Although individual texts can display postmodern
characteristics, many popular music conventions have been misrecog-
nized as postmodern strategies that are typical of music video *per se*
(Frith, 1988; Goodwin, 1993; Straw, 1993).

Although there are clear visual similarities among video dance, music
video and television advertising, it is a matter of debate as to where video
dance is situated in relation to the consumer body. As with music video
and television advertising, video dance is part of a commercial institution that aims to maximize audiences and profits. As with any television text, video dance is subject to marketing decisions and publicity
strategies (Brooks, 1993). It is part of a capitalist system reliant on foreign markets, screen dance festivals and the remuneration of capital
through advertising revenue or television licence sales. Yet there are
also some striking differences. With music video and television advertising, consumerist ideologies are reinforced, and innovative or oppositional discourses are reincorporated into the form. Conversely, video
dance can employ textual radicalism as an end in itself. The actual
commercial potential of video dance is also questionable.

Whereas advertising and music video are pervasive televisual forms,
video dance has had, at present, an extremely limited exposure. There
is a distinct marginalization of dance in general on television, of which
video dance is only a fraction (Allen, 1993). Although *Tights, Camera,
Action! 2* and *Dance for the Camera 2* were programmed at peak times
on a Friday and Saturday evening respectively, the former was screened
on Channel 4 and the latter on BBC2, which are traditionally channels for arts programming and areas outside mainstream interest.

The marginal status of video dance is further supported by audience viewing figures. Such statistics are somewhat crude and must be treated with a degree of critical caution; yet they can offer some idea of trends in audience taste (although this is obviously subject to programming policy). At the *Debate on Dance on Screen Across Europe* (1995), it was revealed that one million viewers tuned in to *Dance for the Camera 2*. In comparison to stage performances, this viewing figure is vast. The Place Theatre, which programmes works from the more experimental end of the dance spectrum (and involves a crossover of certain choreographers and styles of work likely to be seen in video dance), has a seating capacity of 200. This suggests that video dance is reaching viewers beyond the traditional dance audience. Yet compared to other television programmes, this figure is nominal.[7] It begs the question of how effective the promotional discourses of video dance can be when they reach such a minority audience in the first place. The limited commercial viability of video dance is reflected in the continual struggle to raise money in order to fund new video dance projects (Meisner, 1991).

It can be argued that, up to a point, video dance is a discrete phenomenon which displays several features that differ from television advertising and music video. There is clearly an aesthetic overlap among the three forms, but whereas television advertisements and music video are closely tied to a commodity, video dance is not. The promotional potential of video dance is relatively limited in comparison to the ubiquity of advertisements and music videos. Video dance is further distanced from these forms in terms of screening slots. Whereas advertisements are shown during 'commercial breaks' and music video is 'framed' in the commercial context of youth and music television, video dance is screened as an arts programme. Thus, to some extent, video dance resists the promotional agenda of advertising and music video. Yet the placing of the *Paramount Hotel Ads* in a video dance series clearly calls into question the boundaries of 'art' and 'commercialism'.

Video dance undoubtedly shares many striking visual and aesthetic similarities with the texts of advertising and music video. The extent to which these texts draw on and overlap with each other is well documented, from the use of innovative and eye-catching content to the privileging of formal concerns such as framing, composition, lighting and colour. Each genre is characterized by fluid boundaries and thus bodies are constructed in video dance, television advertising and music video that traverse and transgress a multiplicity of aesthetic and commercial

sites. It is apparent that the fluidity of these bodies problematize symbolic boundaries. Notions of 'art' and 'commercialism' are no longer discrete, but are both blurred and disrupted. Irrespective of how these forms
are framed, video dance has the capacity for promotion, and advertisements and music video can be engaged with like art. In order to explore
this aesthetic and conceptual fluidity further, the following section introduces the area of technology in relation to video dance.

Discourses of technology: the technophobic and the technophilic

The video dance body is a body that is technologically mediated. It is
situated at an intersecting point between the medium of dance and the
technology of television, and this convergence suggests that the video
dance body constitutes a hybrid form. It is a technological construction that can only exist within the temporal and spatial frameworks of
television (or film) and cannot exist outside this site. To some extent,
all dancing bodies on television are reliant on technology; however,
whereas television adaptations of stage works attempt to recreate the
characteristics of the live event, video dance sets out to explore and
experiment with the televisual mediation of dance. In video dance, the
televisual technology is inextricably linked to the dancing body in
order to create dance that plays on, utilizes and is determined by the
televisual apparatus.

The late twentieth century has been marked by rapid developments
in technology which have intercepted many areas of social life (Penley
and Ross, 1991; Featherstone and Burrows, 1995). Accordingly, it is not
surprising that technology has also had a considerable impact on the
artistic realm; phenomena such as synthesized music, computer
graphic art and video dance are prime examples of these developments. As McLuhan (1964) famously notes, 'The medium is the message' (p. 7) and this adage appears to be fundamental to any study of
video dance. As televisual technology is central to the genre, this is pertinent to any conceptualization of the video dance body. It is with
these ideas in mind that I want to focus on the concept of 'technology'
as a discursive framework with which to analyse video dance. I therefore aim to explore a number of theoretical discourses ascribed to the
'technological body', in order to push further the idea of a 'fluid video
dance body'.

The 'technologization' of the dancing body by the televisual apparatus
carries certain implications in relation to the performance experience,

spectatorship positions and the dance itself. For instance, the making of dance for television places a whole new set of demands on the performer's body. Whereas the stage dance performance is a single event, the television performance is made up of multiple 'performances', filmed over several days or weeks, which are then edited together into a whole. Each shot can last as little as a few seconds and many shots have to be repeated several times as part of the production process. For the dancing body, the performance experience is fragmented and inconsistent and, once the shooting is complete, there is no opportunity to modify or improve upon its performance. The interaction of technology and dance meanwhile allows the spectator to watch dance in the private setting of the home and to see dancing bodies from multiple viewing positions that could not be accessed in a stage performance. Although the spectator has no choice but to look at where the camera focuses, through devices such as slow motion and close-range filming, the televisual technology can relay facets of movement that are not perceptible when played in real time, or when viewed by the naked eye. The technological mediation of dance has also facilitated new movement possibilities. Bodies can defy gravity, travel at seemingly impossible speeds, repeat identical movement phrases and shoot across the television screen without any apparent physical action or effort on the part of the performer. The technologization of dance through the television medium has clearly opened up new creative possibilities for choreographic practices.

Irrespective of the longstanding history of dance on television and the creative possibilities that television can bring to dance, within dance criticism there is a certain wariness of the interception of technology in video dance. For instance, Bayston (1987) airs a concern that 'there is a danger of the medium becoming the message and the choreography smothered with technology' (p. 707) and Penman (1984) describes *Tights, Camera, Action! 2* as 'dominated by ideology and technology. God help dance if Channel 4 has shown us its future' (p. 1173). There is an almost technophobic sentiment echoed in these fears and concerns, as if the live body were somehow defiled or ruined by technology. Perhaps embedded within the above criticisms is a notion that the live body should not be tampered with, that it is somehow closer to 'nature' and thus preferrential to a body constructed through technology. This negative perspective towards technology immediately reinforces a nature/culture, biology/technology dualism, which is in danger of completely disregarding the symbiotic potential of dance and technology. There is also a notion that the live dance performance is superior to the television performance in that the former is more immediate and carries the 'thrill' of live bodies that are subject to possible risk and failure (Barnes, 1985), in contrast with the latter which is pre-recorded and mediated. Yet, it could be argued that a sense of risk and danger can also be constructed through various televisual devices, such as fast cutting and vertiginous camera angles, and that, with sophisticated filming techniques and improved presentation skills, a televised performance can be as charismatic as a live one.

To some extent the technophobic sentiments towards video dance are irrationally based, because the meeting of technology and dance is not a completely new dimension. A number of technologies are commonplace features of live dance performance. Devices such as electric lighting, electronically recorded and amplified music, and mechanically operated set designs are obvious examples that spring to mind. Choreographers Merce Cunningham and Mark Baldwin have used the computer program *Lifeforms*[8] as a choreographic tool (Brennan, 1996), while other choreographers such as Pina Bausch, Philippe Decouflé, William Forsythe, Daniel Larrieu and Edouard Lock, have all employed film or video within live performance (Bozzini, 1991; Schmidt, 1991).[9] As identified in Chapter 1, there is also a body of dance artists who are currently exploring the potential use of interactive digital technology within stage performance.

Although these technological devices are clearly features of live performance and may affect the way a dancing body is seen, or be a determining factor in the choice of movement that a body has to perform, with the exception of certain digital technologies they are extraneous to the actual body. Yet there are ways in which the live dancing body can be directly subject to technological mediation. For instance, exercise machines are often employed to build up strength in particular areas of the body and contact lenses may be used to enhance a dancer's vision. More drastic measures may include cosmetic surgery as a means to alter the facial features of a dancer; operations that aim to achieve a more ideal dance body by, for example, breaking and resetting the arches of the feet in order to improve a ballet dancer's 'point'; and the use of endoscopic surgery to explore the internal workings of a dancer's body as a way to check out current weaknesses and potential injuries.

To a greater or lesser extent, live stage performances are reliant on a series of technological factors. Yet the video dance body is subject to an additional set of signification and 'texture', which the live body is not.[10] In the case of both dancing bodies, the actual material body, constituted through flesh, blood, bones, and muscles, is the 'raw material' for dance. It is given a sense of 'shape' or 'texture' through a number of formal considerations. These range from the fundamental characteristic of dance, which is dynamic movement in time and space, to considerations such as location, music, costumes, make-up, set design, and lighting. The video dance body is then subject to another set of variables dependent on filmic or televisual codes. Again these involve formal considerations about decisions over camera position, camera movement, lens type, shot length, the use of colour, special effects, and editing choices. This gives the video dance body an additional texture, a layer of 'technological signification'.

The interception of technology in dance carries with it various implications in relation to the dancer's body as a mode of labour, which immediately lends itself to a Marxist analysis. The relations between the labouring body and the means of production are a key issue in Marxist theory. Marx posits that labour is a defining characteristic of human life and that capitalist society has created an alienated labour (Appelbaum, 1988). Although these notions are tied up with the competitive nature of the capitalist market, some result directly from the technological developments of mass production. Within mass production, each worker only plays a small part in the production of a particular product. Appelbaum (1988) states, 'The labor [sic] process is fractionalized and rationalized, with workers reduced to the status of appendages to machines' (p. 73). Consequently, the worker relates to the product that she or he produces as an 'alien object' (McLellan, 1975). As automation has replaced manual workers where possible, the worker is also distanced from fellow workers as mass production puts workers in competition with each other through scarcity of jobs (Appelbaum, 1988).[11]

It could be argued that the concept of the 'alienated worker' has pertinent social and economic implications in relation to video dance. For instance, the way in which the dancing body is subject to technological reconstruction in video dance, perhaps results in an 'alienated body'. As it is the technology that largely facilitates the construction of the video dance body, the dancer's material body only plays a fractional part. In live performance, the material body is the medium that presents the choreography to an audience, whereas in video dance the performance is relayed through the television medium. The dancer's material body is now simply a small part of the televisual apparatus along with lighting, camera angles, location, editing decisions and so on. As with mass production, the individual worker constitutes only an element of the production process, a notion supported by dancer Emma Gladstone's comment about video dance,

> I almost feel like the movement's the least important thing. You know, there's the focus, and the weather and the props and the set and the costumes, and there's so much else that goes into worrying because the medium takes a lot of effort from so many other people to make it happen.[12]

There is clearly a sense of frustration that video dance is so reliant on extraneous technological factors.

Within mass production, where possible, machinery has replaced the labouring body, since increased automation is more effective. This idea of cost efficiency through technological adaptation is one that is pertinent to video dance. This hinges on the notion of whether the technological body, with its potential for simultaneous and multiple performances, is more cost effective than the material dance body, which demands wages and can only facilitate one performance at a time. Although obviously dependent on the scale of the project, some comparisons between the potential needs of and spending on video dance as opposed to stage work can be drawn.

Television is an extremely expensive medium in which to work. For example, a five-minute dance film for the BBC *Dance for the Camera* series is given a £23,000 budget.[13] This not only covers the 'privilege' of using expensive equipment but, in addition to the dancers, also includes a large technical crew made up of camera operators, lighting technicians, sound operators, a design team, make-up artists, a catering crew, a production manager and assistant, plus editing labour and other post-production costs. Yet once a piece of video dance is made its actual performance includes no further costs. A stage performance has to pay for labour and performance costs throughout its running life although it can reap back money on box office returns. Video dance, which makes no direct capital for its producers, may reap nominal and indirect financial returns through the generation of advertising revenue, or through playing a minute part in television licence sales. The only other form of remuneration arises from sales to foreign television channels, but this can never be guaranteed.

In general, it would seem that the 'technological performances' of mainstream narrative films are probably closer to Marx's idea of technology replacing labour. Although feature films are extremely expensive to produce, once in circulation they receive returns, which can be enormous through their ability to access mass audiences. Unlike the stage actor, who can only give a single performance at a time, the recorded performance of the screen actor can reach millions of people at any one time, and the same performance may be seen year after year. Video dance, on the other hand, does not reap such vast audience figures and financial reward, which is reflected in its restricted air time and funding problems (Meisner, 1991). Yet it could be argued that to some extent in video dance, technology replaces labour. The ability of the video dance body to execute simultaneous, multiple performances certainly implies an element of redundancy for the material body. The mobility of the video cassette also allows dance to travel cheaply and quickly to far away destinations independently of the dancing body.

In the light of cultural attitudes towards technology, which provide a fertile ground for debate, the notion of a 'technological body' offers a rich area of research. Its polemical discourses range from the techno-phobic to the technophilic and have been contested from the higher ech-elons of academia (Penley and Ross, 1991; Featherstone and Burrows, 1995) through to the popular representations of film and television.[14] Although society habitually relies on technology, there are deeply embedded attitudes towards technology based on fear and suspicion (Lupton, 1995). The spectrum of ideas that surrounds technology has been echoed through art texts. While certain modernist visual art movements, such as Futurism (Tidsall and Bozzolla, 1977; Benton, 1983), posit a utopian conceptualization of technology, other forms present a more dystopian view. Literary texts such as Mary Shelley's *Frankenstein* (1912) and Phillip K. Dick's *Do Androids Dream of Electric Sheep?* (1968), or cinematic examples such as *The Terminator* (1984) and *RoboCop* (1987), depict representations of technology that have run amok.

The effects of technologies on society and culture have also been addressed by various scholars (McLuhan, 1964; Benjamin, 1973; Baudrillard, 1993; Druckrey et al., 1996) and much of this work points to the implications of technology in relation to art. The mass availabil-ity of the 'screen' and the reproducibility of the 'two-dimensional image' are typical of postmodern culture, and Druckrey (1996) suggests that the 'screen' has become central to the communicative and aes-thetic experience. This is pertinent to the video dance form, which can only be accessed through the television screen. Druckrey describes con-temporary social existence in terms of a 'panoptic culture' in which vision and technology are inextricably linked. Through various record-ing and surveillance devices, the 'gaze' has become associated with

power. Indeed, he even suggests that it is no longer possible to represent or experience the world without technological mediation. This
notion is perhaps supported by Copeland's (1995) observation that live
performance is increasingly mediated by technology.

Comolli (1996) also reinforces the notion of a visual culture mediated by technology. He proposes that, since the mid-nineteenth century, there has been a 'geographical extension' of the visible. He
suggests that as mechanical technology can expose and multiply the
visible, through cinematography the movement of humans has become
increasingly visible. The centrality of movement in relation to the visible is pertinent to video dance, with its focus on motion and its ability
to explore and manipulate the mobile form. Comolli (1996) goes on to
note that the 'mechanical eye' has become superior to the 'human
eye', as the former has the potential to intrigue and fascinate, whereas
the latter is left with limits and doubts. Again, this holds some relevance to video dance in which facets of the moving body are revealed
through the camera that could never be witnessed by the naked eye.

More recently, the 'information revolution', which characterizes the
rapid expansion of digital technologies in everyday life, has been
marked by a proliferation of writings on the subject. 'New technologies' such as personal computers, CD-Roms, the internet, and virtual
reality are becoming increasingly commonplace, and various texts have
sought to examine the social, cultural and economic implications of
contemporary 'technoculture' (Penley and Ross, 1991; Rheingold, 1991;
Crary and Kwinter, 1992; Featherstone and Burrows, 1995; Hables
Gray, 1995). In turn, this is paralleled by a series of art texts. William
Gibson's seminal novel *Neuromancer* (1984) spawned not only a host of
imitators, but a new literary genre known as 'cyberpunk' (Fitting, 1991;
Kellner, 1995; McCarron, 1995). Similarly, a set of performance artists,
such as Franco B, Stelarc and Orlan, are currently involved in work that
explores and examines the application of digital technology in relation
to the 'performing body' (Hungate, 1996).

To examine the phenomenon of the technological body in relation
to video dance, the following two sections address video dance from
the perspective of 'mechanical technology' and 'digital technology'
respectively. Druckrey (1996) argues that whereas 'mechanical culture'
is located in industrialization and modernity, 'computational culture' is
rooted in post-industrialization and postmodernity. The section devoted
to mechanical culture takes Benjamin's (1973) 'The Work of Art in the
Age of Mechanical Reproduction' as its starting-point. Benjamin's
influential essay was written in 1935 and thus reflects certain views

and ideas in currency in the early part of the twentieth century. In contrast, the section on computational and digital culture explores a selection of contemporary writings that focuses on the 'new technologies' of the late twentieth century. This literature cogently captures the cultural *Zeitgeist* with its rapid advancements in technology.

The mechanical body

Benjamin's (1973) seminal essay, 'The Work of Art in the Age of Mechanical Reproduction' focuses on the nature of art in capitalist society. Benjamin's examination of the way in which art has been transformed by technological reproduction makes it a prime site for the discussion of video dance. Benjamin commences the essay by noting that, although all works of art are reproducible to the extent that they can be imitated, mechanical reproduction is different. Reproduced works lack one element that an original work of art retains: a unique existence in time and space. The original work is subject to a history of changes in terms of physical condition and ownership. Benjamin notes that originality is a prerequisite for authenticity, and authenticity exists outside reproducibility. As a result of this, he suggests that an 'original' carries with it a particular 'authority'. Yet Benjamin does not lament this lack of authenticity. He asserts that technical reproduction is independent of the original, a characteristic which he sees as liberating and cathartic (Gasché, 1994). For instance, he suggests that photography is able to bring out elements that the naked eye cannot see and a reproduced copy can be consumed in contexts that would not be possible for the original. This immediately brings to mind an awareness that video dance performances can take place in the home, and perspectives can be shown that could never be achieved in a live performance situation.

Benjamin (1973) conceptualizes authenticity in terms of an 'aura'. 'Auratic art' is tied up with notions of uniqueness, singularity, tradition and continuity. An aura carries with it everything the work has come into contact with through its existence and, although it is beyond appearances, the notion of an aura becomes concrete and real (Gasché, 1994). With reproduced works of art, however, the aura of the original is missing. Benjamin states that reproduction is connected to mass movements and the copy comes to the 'beholder' in his or her context. Yet authenticity is a somewhat problematic notion in relation to dance. With a painting or a sculpture, there is clearly a sense of an original text. This is less concrete, however, in the case of performing arts and particularly dance. Although each stage performance is unique, it is debatable as to whether an 'original' can ever be located.

Due to the ephemeral nature of dance, this begs the question of whether a particular work can ever possess an aura that is carried through time. At least with music and literary-based theatre, an original score or text exists that may be used as a measure of authenticity and, therefore, of authority. Dance, on the other hand, is a transient form and, although a number of elaborate notation systems exists, it is nevertheless extremely difficult to gauge the authenticity of a performance, or designate a specific performance as being the 'original'. For instance, the first performance of *Swan Lake* took place in 1877 and was choreographed by Julius Reisinger, but the work classed as the definitive *Swan Lake* was choreographed in 1895 by Marius Petipa and Lev Ivanov (Koegler, 1982). Most current productions derive from the Petipa–Ivanov version; yet there is no guarantee that contemporary performances bear any resemblance to the original. Although a musical score is subject to contemporary interpretations, there are factors that remain relatively constant such as pitch. Dance, on the other hand, exists on bodies that evolve and deteriorate through time. There is no way that the choreography of *Swan Lake* would look identical on the body of the athletic ballerina of the 1990s to that of the smaller, more shapely dancer of the 1890s.

Yet although authenticity is a particularly slippery concept in relation to dance, nostalgia for the live performance and the authenticity of a unique dancing body form the basis of some arguments that seek to criticize dance on film and television. As highlighted earlier, Barnes (1985) argues that screen dance lacks the sense of risk that live performance has, and that the camera fails to show the unique virtuosity that can come from a particular dancer. This suggests a prioritization for the one-off live performance, even though notions of authenticity are problematic in relation to dance. It is interesting that Benjamin conceptualizes the 'original' in terms of uniqueness and permanence, and the 'copy' in terms of transience and reproducibility. In the case of dance, the live performance may be unique, but it is far from permanent, while video dance aligns itself more with permanence. Although television has been characterized by its transience, in terms of its vast output of fleeting images (Connor, 1989), it could be argued that the video dance performance is permanently recorded on a 'master tape' and, as video recorders are common household objects, television spectators have potential access to their own record of the work, which they may view time and time again. Although Benjamin's argument loses some ground in relation to dance and authenticity, in the case of 'concrete arts' notions of singularity and originality hold some sway.

Benjamin (1973) goes on to assert that as an aura implicitly assumes an original, it is therefore always characterized by 'distance'. Although it may be a short time-scale, the auratic work develops a heritage from its original state through to its current existence. Benjamin suggests that, in contrast, the reproduction eliminates distance. The reproduction can be defined by its immediacy. To some degree this notion is applicable not only to the video dance body, but any body captured on film or video. For instance, if an 'original, live body' performs a movement, from the moment the body executes the movement a distance is created. It will never be possible to go back to that particular movement in time and space again. The 'reproduced, screen body', however, is not characterized by this singularity. Distance is eliminated as it is able to reproduce the movement identically, as an original 'copy' does not exist. The 'reproduced body' is typified through its multiplicity. Thus although it is problematic to define live dance performance in terms of authenticity, it is nevertheless possible to acknowledge an original movement on a live body, because it undergoes a loss of aura in its reproduction.

Benjamin (1973) takes a particularly optimistic stance towards mechanical reproduction, which he describes as being instrumental in the emancipation of art from its ritual context. He argues that an auratic work has a certain 'hold' over the spectator in the sense that it is made distinctive by uniqueness, distance, authenticity and authority (Gasché, 1994); it becomes a 'cult object'. Mechanical reproduction, on the other hand, liberates art from this ritual status. Benjamin (1973) notes:

> for the first time in world history, mechanical reproduction emancipates the work of art from its parasitical dependence on ritual. To an even greater degree the work of art reproduced becomes the work of art designed for reproducibility. (p. 226)

He states that ritual art is tied up with 'cult value'; its existence is more important than being on view. In contrast, he describes reproducible art in terms of 'exhibition value'; it is liberated from ritual as the texts are more mobile and there are multiple opportunities for viewing the work. To some extent this differentiation between cult value and exhibition value may be applied to the contrast between live

stage performance and video dance. For instance, the very first performance of the Petipa and Ivanov *Swan Lake* carries with it a cult value.
The performance can never be reproduced and, consequently, it is
given an almost sacred status. The ballet lives on in the shape of programme notes, reviews, the musical score and the writings or anecdotes of the choreographers, dancers and audience members. The fact
that it can never be exactly reproduced gives it an almost mythical status: it is enough to know that it once existed. With video dance, however, its very function is exhibition. It has no existence outside the
television apparatus. The medium allows video dance to be accessed by
any number of people in the context of their homes and, with the
benefit of video recording, it may be viewed at any particular time.
Distance and singularity are not an issue: the form exists to be seen.

Yet it could also be argued that even reproductions can take on a cult
value through time. As certain reproduced objects cease to be manufac-
tured and become increasingly scarce, they can become collectors'
items and take on their own cult status. In the way that original art
objects can be characterized by a distinct history, in terms of physical
condition and ownership, the same can apply to the reproduced
object. Although the reproduced object will never have the unique status of an original work, the immense cult value of certain reproduced
objects, such as early albums by famous pop groups and old editions of
children's comics, is reflected in the exorbitant prices they achieve in
well-publicized auctions. In decades to come, video dance may take on
the same cult value. As people gradually erase video dance pieces from
their video collections, or as copies become damaged, certain works
may become increasingly valuable. Similarly to an original art work
locked away in a private collection, the 'master copy' of a video dance
work, held in a television company's archive, may be equally inaccessi-
ble. It would seem that certain features of Benjamin's argument are
perhaps more relevant to reproduced objects that are currently in mass
circulation.

Benjamin (1973) raises another issue in relation to technical reproduction which is pertinent to video dance: the difference between
the stage actor and the film actor. The performance of a stage actor is
presented by the actor in person and takes place within the space of a
single performance. The film actor is meanwhile presented by the camera, and the performance consists of many different performances.
Again, Benjamin suggests that the film actor must forgo his or her
aura. There are also many variables involved in making a film that do
not involve the actor such as camera work and editing. Benjamin

describes the film actor's performance as a 'mirror image' that has become separated and transported before a public audience. It is always beyond the actor's reach, which is perhaps cause for anxiety. Many of these concerns are central to the performing body in video dance. Whereas the live dance body has a certain degree of control over its performance, the video dance body is subject to several extraneous factors. This begs the question of whether the dancing body suffers from a certain impotence due to the way in which it is technologically mediated. Whereas a live body can respond to an audience, improve on its performance or break down in some way, the video dance body always executes an identical performance. The lack of opportunity to modify a performance in video dance and the reliance on so many technological factors, such as camera movement, editing devices and lighting, are clearly areas of concern and frustration for the performers involved.

Another feature of mechanical production which has altered the role of the producer/performer of art can be seen in Benjamin's (1973) proposition that due to the pervasiveness of the mass-produced form, 'authorship' has become increasingly available to the everyday public. He uses the example of the press to highlight this phenomenon. Whereas previously the art of writing had been the domain of a small circle of authors, towards the end of the nineteenth century the mass availability of the press has allowed readers to become 'writers', whether in the form of letters to the editor, or reviews and reports printed by professionals from the political, religious, and scientific spheres. Likewise with film, Benjamin suggests that the ubiquity of film news reels has allowed the everyday person to graduate to 'film extra' or 'actor'. This 'democratization' of the arts is equally applicable to television, which draws upon everyday people to become the subject of news stories, quiz shows and documentary programmes. It is perhaps for this reason that in video dance there has also been a considerable number of non-dancers and everyday people. Although there was a period of dance theatre seen in the 1960s Judson Theater collective in New York, in which non-dancers were employed in stage performance, this was very much a temporary rebellion against the virtuosity of previous theatre dance techniques (Banes, 1987). Since that period, there has been a marked return to virtuosity and spectacle in performance. Hence, it could be suggested that whereas 'authenticity' and virtuosity continue to characterize the elite performances of theatre dance, television is marked by its accessibility to non-specialists and everyday people and, accordingly, it is acceptable to utilize people in video dance who are neither trained nor skilled dancers.

Benjamin (1973) asserts that film is socially significant in that its audience is more critical and receptive than the viewer of auratic art. The film apparatus is able to alter fields of perception, analyse events in more depth, reveal new spatial formations, and relay qualities of movement imperceptible to the naked eye. He describes this in terms of 'unconscious optics', which have a certain 'shock value' that results in a critical and distracted audience. Whereas the viewer is 'lured' into the auratic art work, film, he suggests, produces an autonomous spectator. Through 'shock value' the object deflects from itself and prevents the audience from being lost in contemplation (Gasché, 1994). Yet some of these ideas are problematic in relation to contemporary film theory. Many of the aforementioned devices are now regular features of narrative films. Through techniques such as 'invisible editing' the spectator is not distanced from the film, but is said to be 'sutured' into the text (Stam et al., 1992). Recent spectatorship theory proposes that a number of unconscious processes are at work in which the spectator becomes psychically involved with the film (Caughie and Kuhn, 1992). Although this last issue is somewhat debatable, it is probably fair to say that mainstream narrative films have become a highly conventional form and that their 'shock value', at least in formal terms, is negligible. This challenge to Benjamin is supported by Koch (1994), who notes that whereas Benjamin sees the filmic apparatus as being able to 'penetrate reality', in practice, film constructs an ideologically motivated artificiality.

It is arguable that video dance is closer to Benjamin's theory on the production of a distracted, critical viewer. One of the aims of video dance is to explore the formal relationships between dance and the camera, which often results in striking images and peculiar perspectives: bodies can appear to defy gravity, the dance can be shot from unconventional perspectives, cuts occur in unexpected places, and various special effects can construct startling and unusual images. The result is an unpredictable dancing body. This play on conventions possibly has more potential to produce a distanced and distracted spectator position than has contemporary film making, with its well-established codes and conventions, styles and genres.

In retrospect, as views on art have changed, some of Benjamin's ideas perhaps need to be rethought. For instance, it could be argued that, through time, the reproduced object can take on a cult status similar to that of an original art work. It could also be suggested that some of his ideas, in relation to the authenticity of the original art work, fall down when applied to dance due to its essentially transient nature. Yet several, such as the singularity of the original and the multiplicity and accessibility of the reproduction, or his analysis of the fragmented filmic performance, are still currently relevant, particularly in relation to video dance. It is also significant that Benjamin (1973) destabilizes the 'authenticity hierarchy'. He does not prioritize the original art work for its uniqueness and purported authenticity, but rather he values the reproduction for its ability to 'emancipate' art and make it accessible to the masses. His challenge to the primacy of authenticity is perhaps also in line with more recent academic thinking that raises issues about the problematic nature of authenticity. This hinges on the question of whether authenticity can ever be identified and the possibility that concepts of authenticity are subject to social and cultural change. Benjamin's non-elitist and non-technophobic perspective is a valuable advocacy for video dance and other art forms characterized by mass availability, accessibility and mechanical or electronic production.

The digital body

The rapid acceleration and development of new digital technologies have thrown into question existing ideas about the body and its environments (Penley and Ross, 1991; Featherstone and Burrows, 1995; Kellner, 1995), and the impact is documented across a vast array of disciplines that include medical discourses (Hartouni, 1991; Hables Gray, 1995), performance work (Kozel, 1994, 1995; Hungate, 1996), science fiction literature (Fitting, 1991; Kellner, 1995; McCarron, 1995), and academic texts (Haraway, 1991; Crary and Kwinter, 1992). A major effect of these technologies is the part they have played in the creation of new bodies and environments (Kellner, 1995). Through the impact of technology on the body the capacities of the human form have distinctly shifted. Practices such as IVF treatment, cryogenics and genetic engineering offer the body new possibilities in terms of reproduction, mortality, and evolution (Haraway, 1991; Hartouni, 1991; Kellner, 1995). Even the insertion of a pace-maker, which is a standard medical practice nowadays aimed at extending the longevity of the heart, creates a technologically mediated body.

These advances have also posed a challenge to existing notions of space (Featherstone and Burrows, 1995; Kellner, 1995). For instance, computer-simulated space offers new environments for the consumer. An obvious example is the constructed environments of interactive computer games in which the user has to negotiate her or his way through digital space. At the more extreme end of the scale is a virtual reality (VR) environment. Featherstone and Burrows (1995) describe this as, 'a pure information space populated by a range of cybernetic automatons, or data constructs, which provide the operator with a high degree of vividness and total sensory immersion in the artificial environment' (p. 3).

Some areas of cyberpunk literature even go so far as to contemplate a vision of post-bodied, post-human existence (Featherstone and Burrows, 1995). William Gibson's seminal novel *Neuromancer* (1984) depicts a society made up of human-machine hybrids, in which the computer user may leave behind the confines of the human body and its immediate lived space in exchange for a pure data environment known as the 'Matrix'. In the light of recent developments on the internet, Gibson's (1984) vision of the future bears remarkable similarities to the actual present (Fitting, 1991; Kellner, 1995). The corporeal and spatial configurations envisaged by Gibson, and echoed within the technological climate of everyday life, are commonly referred to as 'cyber bodies' and 'cyber space', which derive from the generic term, 'cybernetics' (Kellner, 1995; Tomas, 1995). The term was coined in 1948 by Norbert Weiner, to encapsulate the new ideas that embraced control and communications theories (Featherstone and Burrows, 1995).

This focus on the body and its environments is clearly instrumental not only to dance but also to the more general socio-cultural concepts that shape society's ideas about the body. Indeed, strong metaphorical links between the computer machine and the human form are highlighted through various corporeal and technological discourses. For instance, the human brain is often conceived as an advanced software system (Tomas, 1995), while the computer is believed to possess parallel characteristics with the human body (Lupton, 1995). Lupton (1995) describes the way in which the computer is designated various cultural meanings that are derivative of society's ideas about the human form. For example, users are said to invest emotional relationships in their computers, computer software is susceptible to 'viruses' that are analogous to the HIV virus, and computer networks can be subject to cyber-crimes and 'sexual misconduct'.

The significance of new technologies on the body is dialectically addressed in both theory and practice. Whereas some areas of work pronounce this technological *Zeitgeist* as signifying the ultimate obsolescence of the body, others see the body as a final point of reference for social and cultural signification in this technological age (Kozel, 1994; Featherstone and Burrows, 1995; Hables Gray, 1995). The centrality of the body within these discourses is extremely pertinent to video dance's construction of a technologically mediated body. Although many of the computing technologies employ interactive frameworks, as opposed to the earlier, 'non-interactive' technologies of television and video, there are nevertheless strong parallels between the two bodies as sites of contestation. The heterotopian possibilities that surround the cyber body are perhaps echoed in key ideas in relation to the video dance body. A central polemic to emerge is whether the video dance body, as a technologically mediated form, may constitute a threat to the material body or signify a potential extension of its capabilities. To clarify what is at issue, this section aims to address various similarities, or points of departure, between conceptualizations of the body in digital technology and the video dance body.

Featherstone and Burrows (1995) suggest that the impact of new technologies on everyday existence has created a shift from the public sphere towards an increasingly privatized world. Many facets of social life have been altered to personal consumption, as variously reflected in technologies ranging from miniature television sets and camcorders (Haraway, 1991), through to virtual reality systems and internet access on personal computers (Featherstone and Burrows, 1995). Consequently, new social formations have come to the fore which no longer rely on 'face-to-face' contact, but technologically mediated forms of communication. This is seen in the prevalence of portable phones, faxes and e-mail, which have become essential features of day-to-day living (Featherstone and Burrows, 1995). To some extent this privatization has extended into the arena of dance performance. Whereas previously Western art dance performance took place in the collective setting of the theatre, video dance occurs in the private, domestic locale. Although relatively minimal, there is also an element of interactivity involved in watching video dance that cannot occur in live dance performance. As a work is transmitted the spectator may manipulate the colour, the contrast, the tone and volume of the image, and switch back and forth to other channels to create her or his own relay of images. Although it could be argued that the majority of viewers rarely tamper with their television settings, 'channel zapping' is a regular feature of spectator interaction (Connor, 1989; Allen, 1993). A video recording of a video dance piece permits even further interactivity, allowing the viewer to watch images in fast-forward or to rewind, or pause the image, or to watch individual fragments over a period of time, or repeat certain sections of the video. Although this is nominal compared to the interactivity of computer programs, CD-Roms, and virtual reality environments, the relocation of dance into a private technological setting facilitates certain spectator–text relations that do not exist in live dance performance.[15]

The concept of the 'body' has been radically drawn into question in the light of recent technological discourses, and embedded within these concomitant ideas is a certain dismissal or contempt for the 'material body'. In cyberpunk fiction the actual physical body is described as the 'flesh' or the 'meat' and is conceptualized as being 'heavy' or 'dull'; cyberspace offers the computer user a way to transcend the material body into the realm of 'pure mind' (Fitting, 1991; Balsamo, 1995; Kellner, 1995). Likewise, certain advocates of virtual reality also make claims for the potential to escape the physical body (Rheingold, 1991). Rooted within the epistemological framework of these ideas is a classic Cartesian dualism: a mind–body split (McCarron, 1995; Kozel, 1995). The primacy of experience taking place in a purportedly 'disembodied mind' clearly reasserts this binary opposition. These ideas are also reinforced in 'contemporary cyborg cinema'. Films such as *The Terminator* (1984), *Bladerunner* (1982) and *RoboCop* (1987) construct a dialectical contest between human and machine, and whereas the two may appear identical in bodily appearance, it is the mind that authenticates an entity as being human (Holland, 1995).

This negation and dismissal of the material body is a somewhat alarming concept for the dancing body. Of concern to dance theorists who wish to propound the dancing body in terms of a 'corporeal intelligence' or 'thinking body' (Grau, 1995), rather than as a vessel designed to mediate movement, is the radical reinforcement of the mind–body split. It is perhaps this pervasive Cartesian privileging of the mind that has led to the marginalization of dance within Western culture. Yet it is worth considering for a moment the relationship between technology and the material body in video dance. Although in Chapter 1 Rubidge (1999) allows for a definition of digital dance that does not necessarily display representations of the human or anthropomorphic body, the live body, with its various boundaries and restrictions, is a fundamental starting point for video dance. This is not to suggest for one moment that these boundaries and limitations are fixed. Instead, part of the video dance agenda is to explore the material body and its movement parameters through the televisual apparatus. Yet while it can be achieved technologically, video dance has not sought to construct a body that can do anything. Dislocated body parts and heads that rotate 360 degrees would be nothing more than a technological gimmick, which is perhaps why such devices are generally resisted in video dance. The video dance body is not simply an electronic representation that can mutate in endless configurations. It is a material body that has flaws, quirks, strengths and limitations, but which is open to reproduction, multiplication, distortion and augmentation by the televisual technology in order to become a video dance body.

To some extent the Cartesian dualism embedded within cyberpunk texts is not particularly pertinent to video dance, in the sense that cyberpunk addresses the body of the computer user, while the video dance body constitutes the text, rather than a spectator/user. The former is tied to notions of agency, whereas the latter is simply a representation. This is not to imply that representations cannot signify a mind/body split. The point of this proposition is that it would be unwise to make glib comparisons between the two bodies. Yet the relevance of this theoretical area to dance in general is that the supposed obsolescence of the body in computing and virtual technologies is challenged by several scholars. Lupton (1995) and Balsamo (1995) insist that the user's body remains a material entity and constantly draws attention back to itself through bodily needs such as hunger, thirst, and exhaustion. Kozel (1994) has described the way in which her material body radically called attention to itself in a virtual performance, both as a cultural marker and sensory entity.

Lupton (1995) argues that users have a symbiotic relationship with their computers, and that the centrality of the 'thinking body' within this relationship provides a challenge to the Cartesian dichotomy. For instance, it is the body, and specifically the fingers, which literally activate and control a computer. It is not simply the 'disembodied mind', but a variety of corporeal senses, including the proprioceptive, that are essential to virtual reality (Rheingold, 1991). Lupton (1995) suggests that 'computers may be viewed as contributing to individuals' images and experiences of their selves and their bodies' (p. 99) and describes the way in which computers inscribe the body. She uses the example of the long-term computer user who finds that using a pen feels like an alien activity, drawing interesting comparisons with video dance in which the technology has potentially also shaped the dancer's experience of the body. Video dancers may need to reconstruct facial expressions to suit the intimacy of a camera rather than the distant mass of the audience. As dancer Emma Gladstone explained in Chapter 1, dancers can explore subtle facial expressions and eye movements in video dance that would simply not be perceptible in a stage context. Dancing bodies may become inscribed with alternative patterns of movement as a result of the detailed, gestural vocabularies and pedestrian movement that are often employed in video dance. As a consequence of these ideas, the material body appears to be far from obsolete.

Yet although the material body continues to remain a part of the subject's ontology, it is clear that the body has become increasingly technological (McCarron, 1995; Rawdon Wilson, 1995). This is amplified in cyberpunk, with its prevalence of bodies that have been subject to prosthetic devices, cosmetic surgery and genetic implants (Fitting, 1991; Kellner, 1995; McCarron, 1995); yet it is also reflected in the everyday. Technology is constantly employed to supplement and enhance the capacities of the human body whether in the form of a pocket calculator, a hearing aid or an artificial knee replacement. As McLuhan (1964) notes, technologies are an extension of the body. This merging of technology and the human body has been conceptualized as a 'cybernetic organism', or simply a 'cyborg' (Fitting, 1991; Haraway, 1991; Featherstone and Burrows, 1995; Kellner, 1995; Hables Gray, 1995). The cyborg is described as a human–machine hybrid, which clearly begins to challenge the boundaries of natural and artificial, biological and technological, and human versus machine. These blurred boundaries immediately problematize concepts of identity and subjectivity. If a body is technologically mediated this poses the question of whether it is the same body as the one prior to its reconstruction and where the so-called 'natural body' begins and ends.

Hables Gray (1995) argues that there is no clear-cut definition of a cyborg. Though the term was originally coined by Manfred Clynes and Nathan Kline, who describe cyborgs as 'self regulating man-machine systems' and stress the homeostatic nature of these phenomena (Tomas, 1995, p. 35), it is currently used with greater diversity. The cyborg is usually defined in terms of some degree of merging between, or hybridization of, the organic and the machinic (Hables Gray, 1995); yet conceptualizations range from the everyday to the fantastical. The fictional, high-tech cyborgs of films such as *The Terminator* and *RoboCop* constitute one end of the spectrum. As with the video dance body, these are visual representations of a technologically enhanced body rather than a reality. 'The terminator' is a fictitious character: it is conceived as a human–machine hybrid, a robot encased in human skin, so that its technological constituents far outweigh its organic components (Hables Gray, 1995). Although on the television screen the video dance body is a purely technological representation, this image is constructed through an organic body. Unlike the 'terminator', the video dance body is not regarded as some kind of fantastical techno-robot.

At the other end of the spectrum are 'low-tech' cyborgs, and Hess (1995) proposes that a cyborg constitutes any conflation of the human–machine boundary and any human–machine interface, such as a telephone, a car, a television, or a computer, may be conceptualized as a low-tech cyborg. Indeed, Clynes (cited in Hables Gray, 1995) offers an example of a cyborg in terms of the moment when a human succeeds in riding a bike automatically, so that the bike is almost a part of the human physiology. The same notion may be easily applied to spectacles and walking sticks. It could then be suggested that the moment a dancing body is positioned in front of a camera, this constitutes a human–machine interface and hence the video dance body becomes a low-tech cyborg.

The hybrid status of the cyborg is clearly pertinent to the video dance body, which is a hybrid of dance and televisual technology. The video dance body is also subject to dissolving boundaries between what is natural and artificial, biological and technological in that it can move in ways that challenge the spatio-temporal limitations of the live body. Yet regardless of technology, these dualisms have been challenged by various scholars who see the body as being neither fixed nor natural, but socially and culturally constituted and open to perpetual reconstruction (Shilling, 1993). This would suggest that the 'natural body' is simply a popular mythology. What makes the video dance body similar to the cyborg is the impact of technology on the body. Again, this has strong implications in terms of notions of self-identity. The video dance body can clearly move in ways which it is not possible for the same body to do outside the television medium.

So far as function is concerned, Hables Gray (1995, p. 3) divides cyborg technologies into four types: the 'restorative' cyborg technology is said to reconstruct lost functions (a technological limb replacement is an obvious example of this); a 'normalizing' technology, such as a hearing aid, restores a function back to normality; a 'reconfiguring' technology creates 'posthuman' entities that are different but equal to humans (Hables Gray uses the example of the beings in cyberspace with whom humans interact as an instance of reconfiguration); and the 'enhancing' cyborg technology is the category in which the video dance body most appropriately belongs. Macauley (1995) suggests that one of the interests in recent technological discourses has been the way in which technologies can be used to overcome physical limitation. This is pertinent to video dance in that televisual technology can be employed to enhance the capacities of the dancing body. Yet in video dance the technological enhancement is a temporary phenomenon that can only occur on the television screen. Although the material body is the starting point for any creative investigation, it is the televisual body that is enhanced.

Attitudes towards the cyborg are diverse. In science fiction genres it is often conceptualized as monstrous, and a potential threat to humanity (Holland, 1995; Lupton, 1995). Rooted within these attitudes are both technophobic sentiments and ways of addressing and coming to terms with the accelerated pace of technological society (Holland, 1995). Lupton (1995) describes the way in which the cyborg is imbued with masculine discourses: whereas human skin is a vulnerable and easily broken barrier, the cyborg is constituted by a hard skeleton. The former is prone to breakage and leakage of fluids, whereas the latter is impenetrable. She highlights the way in which these anxieties are gendered in terms of the soft, feminine body and the hard, masculine cyborg. Conversely, Haraway has conceptualized the cyborg within a feminist-socialist framework (Haraway 1991; Penley and Ross, 1991; Lupton, 1995; Tomas, 1995). She locates the cyborg as a potentially oppositional, post-gender entity.

The extent to which a video dance body is a cyborg is debatable. Implicit within the terminology of 'cybernetic organism' are notions of agency and power. González (1995) proposes that 'one can consider any body a cyborg body that is both its own agent and subject to the power of other agencies' (p. 268). Although the video dance body is clearly subject to various technological factors, it possesses no agency in its televisual form. A video dance body is simply a technological representation, a recording that cannot modify its performance in any way. Yet the dissolving boundaries that characterize the cyborg are highly pertinent to video dance. For instance, in *Joan*, the camera is inserted inside the human body to create images of the internal organs. It is a device that immediately poses the question of where the body begins and ends, and disrupts the notion of inside and outside with the skin as a boundary marker. Or else the body can call into question existing spatio-temporal parameters by moving at physically impossible speeds, dancing upside down or dissolving completely, for example.

It is this transgression of the nature/culture, biological/technological dualisms that Haraway (1991) sees as the oppositional potential of the cyborg, and which can allow for post-gendered possibilities (Fitting, 1991; Tomas, 1995). As the cyborg is constituted through 'partial identities' boundaries such as gender are rendered meaningless (Landsberg, 1995; Holland, 1995). Similarly, the video dance body no longer needs to be described in terms of age, ethnicity and gender, but can be characterized by its technological status. Whether it is a nonfully abled body, an ageing body, an obese body or a pedestrian body is arguably irrelevant. The body's 'remarkableness' is dependent on how it is treated technologically. Yet this, and Haraway's theory, are possibly hindered by a Utopian vision. Both the cyborg and the video dance body are grounded in visual representation, and are therefore unavoidably invested in certain social and cultural meanings.

Initially, part of the attraction of computing was its potential to operate as a site of concealment or disguise (Featherstone and Burrows, 1995). The internet is sometimes conceptualized as a gender-free, 'colour-blind' space, in which the user can assume any identity (Balsamo, 1995; Lupton, 1995). Indeed, part of the subversive value of the work of the early computer hackers lies in the shift from the visual aestheticization of the body to a world of numerical abstraction, in which intellectual merit overrides appearance (Clark, 1995). It is this notion that supports the popular mythology of the computer hacker as a pale, spotty, slightly overweight late adolescent, or a young man who is constantly glued to the computer screen (Ross, 1991; Clark, 1995; Lupton, 1995). Yet for all the theorized claims about the potential to transcend the body via digital technology, many of these arguments fall down in practice. One of the immediate discrepancies is the assumption that language is independent of cultural markers such as class, age, gender and ethnicity (Balsamo, 1995). The extent to which the internet user can successfully assume any identity is obviously debatable.

Although written communication is a facet of computing that is arguably independent of visual representation, there are nevertheless many areas of digital technology that do rely on visual imagery. For instance, virtual reality systems require some form of visual representation, and computer programmes employ a graphic element whether it be numerical, verbal or pictorial information. What is particularly interesting is the actual choice of visual representations that are used in digital technologies. Clark (1995) observes that, as the aim of the computer screen is to attract and hold the user's gaze, there has been a return to the prioritization of surface appearance. He goes on to note that the material form of digital imagery is influenced by anterior signifying practices. Many computer-generated images draw on the forms and images of the mass media which are already in constant interplay (Clark, 1995). Consequently, much of the digital technology employs a mimetic component. Human forms have been re-presented as exaggerated bodies grounded in rigid racial and gender stereotypes (Balsamo, 1995; Hess, 1995; Holland, 1995). Hables Gray (1995) also argues that cyborgs are closely tied to embodied notions of class, race and gender. Although Clark (1995) moots the possibility of multi-gendered bodies and entities that evolve their own forms, appearances and behaviour in cyberspace, Balsamo (1995) suggests that the reason for the prevalence of stereotypical bodies is that when one set of boundaries is challenged in some way, then another set becomes more rigid. As technology represents a threat to the boundaries of human and machine, nature and artificiality, categories of gender and race are none the less radically reinforced. Constructs of the body within new technologies are perhaps not as subversive as they may at first seem.

Although there are certain aspects of the video dance body that share key features with the cyborg, such as the use of technology to enhance the capabilities of the human body and the blurring of boundaries between biology and technology, there are also points of departure. For instance, the video dance body has no agency: it is simply a pre-recorded representation. It is also significant that, unlike the cyborg, at no moment does the material body actually merge with technology. An image of the material body is modified, distorted and controlled by technology, but these can never be permanent characteristics. The video dance body is never a self-sufficient body that can achieve these technological feats independently of the television screen. The techno-logical modifications are temporary and the actual material body is always absent from the technological mediation. Although it is not wholly appropriate to conceptualize the video dance body as a cyborg, in recognition of the close relations between technology and the video dance body it is perhaps feasible to regard the body in video dance as a 'prosthetic techno-body'.

Prosthetic body parts are a regular feature in cyberpunk literature (Kellner, 1995; McCarron, 1995). Rawdon Wilson (1995) suggests that, whereas a cyborg is an integrated being, a prosthesis is simply a supplement that can amplify and extend the capacities of the material body. He suggests that the prosthesis supports rather than replaces, and defines a prosthesis as, 'an artificial body part that supplements the body, but a part that carries an operating system different from the body's organic processes' (p. 243).

To some extent this is applicable to the video dance body. Although the televisual technology is not integrated with the dancing body so that it becomes a permanently technological entity, the televisual apparatus clearly supplements the body in order to extend and amplify its movement capacity. The televisual technology could thus be conceptualized as a prosthesis to the dancing body. It is through this pros-thesis that the live dancing body becomes a video dance body. Rawdon Wilson (1995) propounds that the notion of prosthetic enhancement is accompanied by contradictory attitudes based on yearning and disgust. While on the one hand there is a desire for technological enhancement, the prosthesis also problematizes the boundaries of the body and notions of identity which produce sentiments of disgust (Rawdon Wilson, 1995). This bias against the technological supplement is perhaps reflected in those critical writings that either denounce or overlook the intrinsic role that technology plays in the construction of the video dance body.

Although there are deeply embedded negative cultural attitudes towards technology, technology is inextricably linked to the video dance form. Televisual technology is employed to modify, distort and enhance the capacities of the live dancing body in order to create dance that could not exist outside the television screen. The technologization of the dancing body carries implications for the choreographer, the spectator and the performer. The relationship between television and dance clearly opens up new choreographic possibilities. The relocation of dance into the domestic television context allows for an element of spectator interactivity and a multitude of spectator positions. In creating video dance the performer has a completely different experience from the stage dance performance. The body itself is also more versatile: it can transgress the limitations of the material body to become a more 'adaptable' or 'fluid' body. The video dance body does not conform to the fully integrated, self-sufficient system of the cyborg model; yet the technological supplementation of the body in video dance does draw into question existing corporeal boundaries in terms of certain dualist frameworks such as biology and technology, nature and artificiality. The fact that the video dance body is technologically enhanced raises questions about where 'biology' and 'technology' begin and end. These shifting and unstable boundaries highlight an element of subversion and 'fluidity' in the theoretical and material construct of the video dance body.

The fluid body: transgression and disruption

The notion of a 'fluid body' has been a recurring theme throughout this chapter. To commence, the concept of 'fluidity' lends itself well to dance; as opposed to stasis and fixity, the dancing body is characterized by mobility, dynamics and shifting configurations. Fluidity is also a key concern in relation to recent theories of the body (Shilling, 1993). This area of work conceptualizes the body as a plastic and unstable phenomenon which is open to perpetual reconstruction. Yet the idea of a fluid body is particularly pertinent to the video dance body as a hybrid entity. The video dance body is not singular but multiple, inscribed with discourses from the fields of dance, television and other visual, kinetic and technological practices. At the beginning of the chapter it is stated that video dance can be defined as a hybrid site. The concept of hybridity stems from the way in which video dance is constructed through two distinct sites: a fusion of a postmodern stage dance tradition with the televisual apparatus. Indeed, it could be argued that video dance straddles a number of different material sites and theoretical disciplines and the way in which it slips in between them suggests fluidity. Although the video dance body traverses different aesthetic practices and intellectual frameworks, it has an uneasy relationship or creates an element of disruption within these sites.

Whereas some forms of screen dance minimize or 'camouflage' the televisual apparatus, video dance calls attention to it in order to create potent visual imagery. Televisual devices act upon dance in such a way that this construction of dancing bodies could not be achieved on stage: bodies are fragmented, they defy gravity, are seen from distorted and unconventional perspectives, and challenge spatial and temporal logic. It is not just the physical body that 'dances', for the camera and the edit are also inextricably bound up with the actual dance. Without them the dance would no longer be the same. The body is 'televisualized', which results in the construction of a video dance body that exposes and plays on the televisual apparatus through employing it in challenging and innovative ways.

The televisual mediation of dance creates a 'video dance body' that transcends the limitations of the material body and which offers the possibility of alternative modes of dance. The intervention of technology has clearly opened up new choreographic possibilities as the spatio-temporal boundaries of the body can be made to appear increasingly fluid, and dynamics can be manipulated independently of the physical body. Since the television context differs radically from the live theatre setting, the choice and design of movement is inextricably linked to the televisual apparatus. Video dance clearly undermines critics' notions of what 'dance' is, with its use of small and detailed action that complements the close-up of the camera; simple geometric lines and everyday movement behaviour that can be accommodated on the small television screen; bodies that are televisually enhanced to create movement that cannot be achieved on stage; and movement that is determined by geographical location. Although this may not be dance in any conventional sense, it can be argued that video dance derives from the triadic relationship between the motion of the physical body, the camera and the cut. Thus definitions of 'dance' and the critical apparatus which surrounds it are challenged and displaced.

The hybrid status of video dance results from a dialectical process in which the televisual apparatus creates dancing bodies that could not be achieved on stage, and postmodern stage dance strategies are relocated to the television context. Drawing on McCabe's (1981) notion of the 'classic realist text', it can be argued that video dance resists the realist devices that pervade the television network through exposing and subverting televisual conventions. This challenge to realist practices results partly from the experimental agenda of video dance that seeks to explore and challenge aesthetic conventions, but also from the postmodern stage dance context out of which video dance has emerged. A deconstruction of hegemonic apparatus and a playful and subversive treatment of established devices are key features of postmodern stage dance.

Whereas conventional television texts are characterized by the linear and logical progression and the spatial and temporal coherence of the classic narrative system, video dance employs fragmented and aleatory structuring devices. The illusionistic, psychologically motivated characters of television realism are similarly absent from video dance. Instead, video dance employs the 'performing body' in alternative ways: in some cases it plays with formal devices, so that the body is constructed as a mobile visual design rather than as a 'character' or 'person', or else it treats the subject of 'human affairs' through multi-faceted representations that are contradictory and unstable. This suggests that the performers of video dance can create fluid and nomadic identities. Although television is a visual medium, due to its relatively poor quality image it tends to rely on its audio component and the transmission of verbal information. In order to draw attention back to the visual element, video dance employs bold colours, striking costumes, alternative locations and peculiar circumstances and events to create arresting and unconventional images. With video dance, the television context is treated as a site of artistic exploration and innovation, an area of television that is rarely explored.

Video dance clearly challenges the format of television in terms of its disruption of realist practices, its break with linear narratives, its refusal to conform to conventions of 'character', and its dominance of visual aesthetics over verbal information. The sense of disruption may be explained by the hybrid status of video dance; the merging of postmodern dance with televisual practices results in conventions being employed and abandoned from both fields. This would suggest an artistic tension in which each discipline attempts to act on the other. In some video dance works the televisual codes take a predominant role and in other instances postmodern dance practices come to the fore. In either case, video dance reveals a dialectical contest between two distinct disciplines that culminates in the creation of a new artistic form. Video dance conforms neither to dance nor television; instead it is a discrete art in its own right.

In addition to the discourses of dance and television, video dance transgresses into the conceptual and aesthetic domains of promotion and technology. In the first part of this final chapter it is suggested that there are stylistic similarities among video dance, television advertising and music video in that they all seek to attract and hold the viewer's gaze through eye-catching imagery. This can be described in terms of 'leaky boundaries' in the sense that each form is constantly reproducing and appropriating ideas from the rich network of visual images that pervade late twentieth century/turn of the millennium society. Consequently, this fluidity calls the boundaries of art and commercialism into question: advertisements can be created that are aesthetically innovative, while video dance can fulfil a promotional function. Yet the commercial potential of video dance is extremely limited: unlike advertising and music video, it is neither framed as a commercial product nor employs devices that exploit financial possibilities. In turn, the experimental and eye-catching images of advertising constantly refer back to a product and the striking visual messages of music video allude to a 'single' and any related merchandise of the band.

The extent to which video dance can be characterized as promotional, and advertising and music video as artistic, largely depends on how the reader or producer frames the text. For example, a spectator may refuse to take on board the selling message of an advertisement or music video and only engage with its aesthetic components, in the same way that a choreographer's motivating factor for making a piece of video dance might be that it has the potential to reach a larger audience than stage dance. Although this issue remains unresolvable, one proposition that does hold some sway is that the similarity between video dance and the other two forms is due to its hybrid status. As dance is being conceived for the television context, it draws on many shared conventions of television imagery and consequently overlaps with the visual components of advertising and music video. Stage dance, on the other hand, has not generally appropriated commercial imagery in this way, except for a handful of practitioners who have an interest in popular forms (Bozzini, 1991; Briginshaw, 1995–96). This fluidity highlights the way in which video dance has traversed and disrupted the boundaries of art and commercialism.

In the second part of this chapter it is proposed that video dance is a technologically mediated art form, and scholarly literature that deals with the relationship between technology, the body and art provides a pertinent avenue of research with regard to the technologizing of dance. Although some of Benjamin's (1973) ideas are limited in their scope, he nevertheless offers a cogent analysis of the effects that technology can have on art, which are germane to the spectatorship experience and the 'performing body' in video dance. The interception of televisual technology allows the spectator to have a greater element of interaction, to take up a multitude of viewing positions that could not be replicated on stage, and to see facets of dance that could not be perceived by the naked eye. Video dance has also placed new demands on the dancing body. On the one hand, the performance is a fragmented experience that, once recorded, allows no room for improvement or modification; on the other, any mistakes can be edited out of the final work so that the 'perfect' performance is always achieved. To a certain extent, video dance prompts a redundancy of the material body. Once the filming process is complete, the live body is no longer required as the screen body continues to execute identical and simultaneous performances. The live dancing body is meanwhile becoming inscribed with video dance practices. The video dancer is often required to perform movement that would not be suitable for stage or to make physical adjustments to suit the specifics of the televisual medium. Video dance has also democratized the dancing body in the sense that performers do not have to be a particular age, shape, build or weight to achieve the technologically mediated movement that is characteristic of video dance. Aesthetic innovation supersedes physical virtuosity.

Whereas Benjamin (1973) is concerned with how mechanical reproduction liberates art, writings on digital technologies focus on the enhancing possibilities that technology can bring to the body. As is the case with numerous digital technologies, televisual technology can enhance and extend the capacities of the dancing body. The video dance body does not, however, conform to the fully integrated model of the cyborg, with its own sense of agency; its technological supplementation is never permanent and video dance continues to use material bodies that have their own flaws, strengths, quirks and limitations. Yet in recognition that the video dance body is technologically enhanced, it can be characterized as a 'prosthetic techno body' to account for its temporary supplementation. Since the video dance body is able to transcend the capabilities of the live body , the dualisms of nature and artifice, and biology and technology are called into question. The video dance body can 'do' the unnatural and exceed biological limitation. Hence it appears that video dance constructs a more versatile or 'fluid' body than the live one, in that physical boundaries are surpassed.

In conclusion, it can be argued that due to the hybrid status of video dance, a fluid body is constructed which has the potential to transgress and disrupt symbolic boundaries. Video dance is marked by a creative tension as a consequence of the competing practices of dance, television and other visual and kinetic forms. This playful manipulation and disruption of existing boundaries has implications for the dancing body, choreographic methods, the spectatorship experience and television as a context for dance. Because of its hybrid character it is essential that any study of video dance employs an interdisciplinary methodology. Whether in examining the role of the dancer or the historical development of video dance, the symbiotic relationship of dance and television is a determining factor of any conclusions that may be drawn. Although this book has touched on the overlap between video dance and advertising, music video and both mechanical and digital technologies, there are clearly further avenues for research. Other potential lines of enquiry that immediately spring to mind are the links between video dance and avant-garde film, video art, dance photography and postmodern television texts. I hope, however, that this book has made a vital start in probing the potential knowledge that can be gleaned from video dance through opening up the debate as to its character, based on the concept of 'hybrid sites and fluid bodies', and through examining the implications this has for dance theory and practice. The direction of future scholarly research is largely dependent on the issues and problematics that video dance throws up in the new millennium. To recall director David Hinton's words, '[video dance] is an idea, the potential of which has hardly begun to be explored'.

Appendix: List of Works Discussed in Chapters 3 and 4

Tights, Camera, Action! 1 (1993)

Topic II/46 BIS Choreographer: Sarah Denizot, Director: Pascale Baes
Tango Choreographer/Director: Zbigniew Rybczinski
Codex Choreographer/Director: Philippe Decouflé
Flesh and Blood Choreographer: Lea Anderson, Director: Margaret Williams
KOK Choreographer/Director: Régine Chopinot
Kissy Suzuki Suck Choreographer/Director: Alison Murray
Perfect Moment Choreographer: Lea Anderson, Director: Margaret Williams

Tights, Camera, Action! 2 (1994)

La La La Duo No 1 Choreographer: Edouard Lock, Director: Bernard Ebert
L'Envol de Lilith Choreographer/Director: Cécile Proust and Jacques Hoepffner
Mothers and Daughters Choreographer: Victoria Marks, Director: Margaret Williams
Monologue Choreographer: Anna Teresa de Keersmaeker, Director: Walter Verdin
Keshioko Choreographer/Director: Saburo Teshigawara
Paramount Hotel Ads Choreographer/Director: Pascale Baes
Joan Choreographer: Lea Anderson, Director: Margaret Williams
Bruce McLean Choreographer: Bruce McLean, Director: Jane Thorburn
3rd Movement Choreographer: Matthew Hawkins, Director: Deborah May
Waiting Choreographer/Director: Lea Anderson
Elegy Choreographer: Douglas Wright, Director: Chris Graves
Le P'tit Bal Perdu Choreographer/Director: Philippe Decouflé
Relatives Choreographer: Ishmael Houston-Jones, Director: Julie Dash
Le Spectre de la Rose Choreographer: Lea Anderson, Director: Margaret Williams

Dance for the Camera 2 (1995)

Alistair Fish Choreographer: Aletta Collins, Director: Tom Cairns
Outside In Choreographer: Victoria Marks, Director: Margaret Williams
Touched Choreographer: Wendy Houstoun, Director: David Hinton
Drip Choreographer: Matthew Bourne, Director: Frances Dickenson

Dance for the Camera 3 (1996)

Dwell Time Choreographer: Siobhan Davies, Director: David Buckland
Horseplay Choreographer/Director: Alison Murray
T-Dance Choreographer: Terry John Bates, Director: John Davies

boy Choreographer: Rosemary Lee, Director: Peter Anderson
Pace Choreographer: Marisa Zanotti, Director: Katrina McPherson
Never Say Die Choreographer/Director: Nigel Charnock
Echo Choreographer: Mark Baldwin, Director: Ross MacGibbon
Storm Choreographer: Aletta Collins, Director: Tom Cairns
Man Act Choreographers: Man Act, Director: Mike Stubbs
Hands Choreographer: Jonathan Burrows, Director: Adam Roberts
Cover-up Choreographer: Victoria Marks, Director: Margaret Williams
Attitude Choreographers: RJC Dance Theatre, Director: Anne Parouty

Notes

1 Dance on Screen: A Contextual Framework

1. Van Manen had originally intended to screen a live recording of the performance, but at that time the playback image was not of a good enough quality. In 1979, however, with *Live*, van Manen was able to achieve the simultaneous transmission of a live performance alongside a live recording. This set a new standard for innovation and technical competence with regards to the use of video in stage performance (Schmidt, 1991).
2. A take is 'a version of a shot. A filmmaker shoots one or more takes of each shot [but] only one of each group of takes appears in the final film' (Monaco, 1981, p. 456).
3. Pritchard (1985–96) notes that the term 'silent screen' is something of a misnomer in that many of the early dance films were accompanied by live music or gramophone recordings.
4. A 'tracking shot' is 'any shot in which the camera moves from one point to another either sideways, in, or out. The camera can be mounted on a set of wheels that move on TRACKS or on a rubber-tired DOLLY' (Monaco, 1981, p. 458).
5. Feminist critiques of Berkeley's work comment on his objectification of women through the cinematic apparatus (Fischer, 1986).
6. A 'reaction shot' is a 'shot that cuts away from the main scene or speaker in order to show a character's reaction to it' (Monaco, 1981, p. 449).
7. See Rowson Davies (1982–83) and Penman (1986).
8. The following articles address some of the issues around filming dance for television: Brooks (1987–88), Lockyer (1983), Merrett (1992), Nagrin (1988) and Nears (1987).
9. See Constanti (1988), Pascal (1988) and Rubidge (1988c).
10. The model of screening short dance works designed specifically for television can also be seen with the *4 Tokens II* (1995) series in Holland, *Dances for a Small Screen* (1997) in Canada and *Microdance* (1998) in Australia.
11. There are, however, limitations associated with the programme: in previous versions the digital figures did not have articulated spines and certain physical details were absent, such as facial features and digits (Jones, 1996). Additionally, the gravitational pull and movement dynamics that characterize live moving bodies could not be replicated or manipulated. Although a new programme has been developed in which the figures have semi-articulated spines and more precise physical detail, the problems with body weight and dynamics remain.
12. See *Hands* (1996, choreographed by Jonathan Burrows and directed by Adam Roberts), *Outside In* (1994, choreographed by Victoria Marks and directed by Margaret Williams) and *Tango* (1980, directed by Zbigniew Rybczynski).
13. A 'point of view' shot is one which shows a scene from a particular character's perspective.

14. Practitioners regularly attend the *IMZ Dance Screen* festival and other dance festivals, such as *Dance on Screen* at the Place Theatre and the South Bank's *Ballroom Blitz*, to show and discuss their work in screen dance.

15. See Mackrell (1988), Merrett (1992), and Rubidge (1988c) as examples of the few articles that do give a voice to the practitioners involved in making dance for the camera.

16. Margaret Williams is the series director for *Tights, Camera, Action!*. In collaboration with choreographer Lea Anderson, Williams directed *Joan* (1994) and *Le Spectre de la Rose* (1994) for *Tights, Camera, Action! 2* and in collaboration with choreographer Victoria Marks directed *Outside In* (1994) for *Dance for the Camera 2*, *Cover-up* (1995) for *Dance for the Camera 3*, and *Mothers and Daughters* (1994) for *Tights, Camera, Action! 2*. Author's interview with Margaret Williams, 5 August 1996.

17. Tom Cairns, originally a set designer, directed *Alistair Fish* (1994) for *Dance for the Camera 2*, and *The Storm* (1995) for *Dance for the Camera 3*, in collaboration with choreographer Aletta Collins. Author's interview with Tom Cairns, 8 October 1996.

18. David Hinton initially worked as a director for the arts programme, the *South Bank Show*, where he made documentaries on Siobhan Davies and Karole Armitage. He has also been closely involved with DV8 Physical Theatre and adapted two of their stage works for screen: *Dead Dreams of Monochrome Men* (1988) and *Strange Fish* (1993). In 1994, in collaboration with choreographer Wendy Houstoun, he directed *Touched* for *Dance for the Camera 2*. Author's interview with David Hinton, 10 October 1996.

19. Alison Murray both directs and choreographs her own work. This includes *Kissy Suzuki Suck* (1992) for *Tights, Camera, Action! 1*, and *Horseplay* (1995) for *Dance for the Camera 3*. With *Wank Stallions* (1993) she won the 'Best Experimental Work' category at the IMZ Dance Screen festival in 1993. Author's interview with Alison Murray, 7 October 1996.

20. Wendy Houstoun choreographed *Touched* (1994), but has also worked with DV8 Physical Theatre as a performer and with David Hinton on *Strange Fish* (1993). Author's interview with Wendy Houstoun, 17 August 1996.

21. Dancer Emma Gladstone performed in *Touched* (1994) and has worked on several dance films by Lea Anderson and Margaret Williams including *Cross Channel* (1992), *Flesh and Blood* (1992) and *Perfect Moment* (1992). Author's interview with Emma Gladstone, 17 July 1996.

22. Anna Pons Carrera performed in *Mothers and Daughters* for *Tights, Camera, Action! 2*, but has also recently worked with Lea Anderson on both a live version and a television reworking of Anderson's *Car* (1996). Author's interview with Anna Pons Carrera, 22 August 1996.

23. *Sardinas* (1990) is filmed using a single top shot as the dancers roll and wriggle across the floor, but on the upright television screen the movement appears weightless and two-dimensional.

24. Although there are several commonalities between film and television, for a detailed discussion of the specificities of film see Monaco (1981).

25. The obvious exceptions to a rectangular stage space are 'theatre in the round', and 'site specific works', which employ unusual locations.

26. A full shot is one of an entire body within a tight frame. This is slightly different from a long shot, which may show a whole body but can also be surrounded by a vast array of background scenery.

27. The term 'editing' is generally employed by the American and UK film industries, whereas 'montage' tends to be used within European cinema. Montage can also be employed in a more specific sense to describe the cinematic device established by Sergei Eisenstein in which 'adjacent shots should relate to each other in such a way that A and B combine to produce another meaning, C, which is not actually recorded on the film' (Monaco, 1981, p. 442).

2 Images of Dance in the Screen Media

1. It is noted in Chapter 1 that *Flashdance* is actually a film, although it has been regularly screened on television. In recognition of its cinematic origins, throughout this section I will refer to it in terms of 'filmic apparatus'.
2. The construction of the body as provocatively sexual is also apparent in *Saturday Night Fever* and *Dirty Dancing* in which the male protagonists use their dancing bodies to incite a sexual relationship with a female character.
3. The diegesis is 'the denotative material of film narrative, it includes, according to Metz, not only the narrative itself, but also the fictional space and time dimensions implied by the narrative' (Monaco, 1981, p. 428).
4. The term 'nude' is used in preference to 'naked'; whereas naked simply means to be without clothes, nude implies a certain constructedness and the embodiment of a way of seeing (Berger, 1972).
5. This idea is clearly reflected in *Fame* in which, early on in the film, the dance teacher warns the students that 'fame costs' and that they have to start 'paying for it in sweat'.
6. The rehearsal scenario is one that comes up in a number of dance films, including *Fame, Footloose* and *Dirty Dancing*, which depicts the time and effort that a dancer must put into training and rehearsing.
7. 'Transformation' is a fundamental part of 1980s dance film, which follows the narrative shift from a rehearsing/training/aspiring dancer to a 'fully fledged dancer'. This is apparent in *Flashdance* in which Alex rehearses and aspires to become a ballet dancer, and is then transformed as she succeeds to impress the audition panel to win a place at the much coveted ballet school. The transformation theme is also apparent in *Fame* in which the students move from daily class through to their end-of-year graduation where they are now fully trained dancers.
8. The text of an advertisement is known as the 'copy'.
9. In addition to the singers on the sound track singing the words 'Just take five', the piece of music that makes up the sound track is actually called *Take Five*.
10. Images of dance have also featured in television advertisements for Maltesers, Snickers and Time Out chocolate products.
11. The name of this particular Twirl campaign is 'Fame', which suggests a conscious reference to 1980s dance film on the part of the advertising agency.
12. This device of linking the colours of the product with the colours of the setting or the performers' clothes has been described as an 'objective correlative' and is a common device within advertising (Dyer, 1988).

13. For accounts of the historical developments of music video see Frith (1988), Kaplan (1988b), Goodwin (1993) and Straw (1993).

14. This may be because much of the analysis has derived from the fields of psychoanalytic, feminist, film and postmodern theory, which are primarily concerned with the image and thus neglect the popular music context out of which music video has emerged (Goodwin, 1993).

15. See Goodwin (1993) for a detailed study of music video examples in relation to his concept of visualizing sound.

16. These ideas are very much in relation to conventional approaches to filming full-length ballets. In January 1996 a series of three programmes was broadcast by the BBC titled *Evidentia* (presented by Sylvie Guillem, produced by Bob Lockyer), which attempted to explore new relationships between the ballet body and the camera.

17. Although both LCDT and Rambert Dance Company had been screened on television prior to these works, this tended to be in the form of documentaries, excerpts and lecture-demonstrations.

18. The only exception to this is in *Cell* in which a number of unconventional camera perspectives are employed but, as stated earlier, they are closely tied to the psychological content of the piece.

19. Although many dance pieces were made for the camera prior to 1956 (Rowson Davis, 1982–83), the majority of these screenings were not recorded and consequently do not exist in the BBC Television Film and Videotape Library. The ballet *Contrasts* from *We are your Servants* is one of the first works commissioned for television that remains in the BBC archives. *Houseparty*, a more substantial television ballet from 1964, appears in the *Catalogue of Ballet and Contemporary Dance in the BBC Film and Television Library 1937–1984* (Penman, 1986) as being the next major commission for the BBC.

20. The reason for this close-up may largely be to do with the fact that it focuses on Jennifer Gay who was formerly a child presenter with BBC Children's Television. Prior to the ballet, Gay is interviewed by Huw Wheldon and hence the close-up may simply be a reminder of Gay's 'success'.

21. Panning is the 'movement of the camera from left to right or right to left around the imaginary vertical axis that runs through the camera' (Monaco, 1981, p. 444).

3   Video Dance: Televisualizing the Dancing Body

1. The term 'video dance' is widely used in dance media circles to refer to this new genre of screen dance, although in screen dance festivals, such as the *IMZ Dance Screen*, this category has been given other titles such as 'screen choreography'. For the purpose of this book the term video dance is used throughout. This is partly for reasons of consistency and continuity, but the emphasis on the word 'video' is useful in that all of the works under examination were screened on television and the television context is vital to the analysis.

2. The two most established festivals are the *IMZ Dance Screen* (Jordan, 1992; Burnside, 1994) and the *Grand Prix Vidéo* Danse (de Marigny, 1991; Chaurand, 1993), which both operate as competitions. There are also a number of smaller festivals such as *Canal Dansa* in Spain, *Dance on Screen* in the United Kingdom, *Springdance Cinema* in the Netherlands and *Moving Pictures* in Canada.

3. Anecdotal evidence from screen dance festivals would also suggest that the differentiation between 'faithful stage adaptations' and 'reworkings of stage choreography' is ill defined, which often leads to some confusion over which category a work should belong.

4. A full list of works that were screened on *Dance for the Camera* (2 and 3) and *Tights, Camera, Action!* (1 and 2) is provided in the Appendix.

5. In the production of any images, the potential mode of reception is funda mental to both the creative process and spectator–text relations. Hence an analysis of video dance within a 'dance screen festival' may produce a different set of conclusions from an examination of video dance within a domestic viewing context.

6. The only exception is the video dance piece *Perfect Moment*, which took up a whole episode.

7. A 'dissolve' is 'the superimposition of a FADE OUT over a FADE IN' (Monaco, 1981, p. 429). A 'fade out' is when an image slowly disappears and a 'fade in' is when an image gradually appears (Monaco, 1981).

8. In film and television, 'classic realism' is a device that attempts to create an illusion of 'reality'. For this reason *Alistair Fish* employs naturalistic or everyday movement. A detailed discussion of 'classic realism' is provided in Chapter 4.

9. 'Pixillation' is 'a technique of ANIMATION in which real objects, people, or events are photographed in such a way that the illusion of continuous, real movement is broken, either by photographing one frame at a time or later printing only selected frames from the continuously-exposed negative' (Monaco, 1981, p. 446).

10. This quote is taken from the author's interview with Alison Murray that took place on 7 October 1996.

11. Although in the 1960s the Judson Church Collective, New York, experimented with the use of 'everyday movement' in order to challenge existing preconceptions of what characterizes dance (Banes, 1987), the use of everyday movement within video dance has nevertheless instigated several British dance critics to question whether or not such movement constitutes dance (Bayston, 1992; Penman, 1994).

12. The 'steadicam' is a camera mount device that can be attached to the camera operator and which 'permits hand-held filming with an image steadiness comparable to TRACKING SHOTS' (Monaco, 1981, p. 454).

4 Postmodern Dance Strategies on Television

1. McCabe's (1981) concept of the classic realist text is examined in further detail in the following section, 'Breaking the Realist Code'.

2. Although McCabe's concept of the classic realist text was developed in relation to the nineteenth-century novel and cinema, it is also applicable to television. See Fiske (1989) and Harris (1996).

3. As with postmodernism, the debate as to what constitutes modernism across various art forms is equally complex and problematic. One feature, however, that has come to characterize certain modernist works is an exploration of the aesthetic and formal features of a particular art form. As Rubidge (1984) suggests that television rarely sets out to explore its own formal features, and Wyver (1986) notes that there is no modernist tradition in television, it would seem more appropriate to employ the notion of a 'classic television framework' rather than a concept of 'modernist television'.

4. It should be noted that postmodern characterizations of music video and *Miami Vice* have been challenged by several theorists. See Goodwin (1993) and Kellner (1995).

5. Of course, there is also a counter-argument in relation to dance that suggests, because dance is often seen quintessentially as the most 'natural' or 'primitive' of the arts, in its use of the body and 'gesture' it has a closer relationship with 'reality' than the other arts.

6. The photograph is clearly a still image and is achieved by a different technological process from film and television. The debate, however, that surrounds the relationship between photography and reality is nevertheless appropriate to film and television theory. As with photography, film and television also use technological apparatuses designed to reproduce images of objects and events from 'everyday reality'.

7. 'Depth of field' is 'the range of distances from the camera at which the subject is acceptably sharp' (Monaco, 1981, p. 428) and 'deep focus cinematography' is 'a technical device which enables film-makers to show foreground, middle ground and background simultaneously in one shot with equal clarity' (Williams, 1980, p. 36).

8. 'Eye line match' devices, point of view and 'reverse field' shot structures are filmic and televisual conventions that situate the spectator within the events of the narrative and the system of 'dramatic looks'. These techniques are particularly evident in dialogue scenes between two characters in which the camera cross-cuts from one actor to the other. The position of the camera is consistent with the 'eye line' of each character and as the on-screen character speaks, the spectator sees that character from the off-screen character's 'point of view'. This cutting back and forth between two characters in dialogue is typical of 'field/reverse field' shot structures (Fiske, 1989).

9. This theoretical position has been criticized as it positions the spectator in such a way that it does not allow for alternative or resistant readings (Fiske, 1989; Harris, 1996).

10. The term 'suture' means 'stitching' and refers, first, to the way in which the shots in a film or programme are 'stitched' together to give an illusion of 'seamlessness' and, second, to the way in which the spectator is stitched into the narrative through such devices as point of view editing and 'reverse shot' structures (Fiske, 1989).

11. The term 'offer' is used as it acknowledges the potential for the viewer to take up alternative or resistant spectatorship positions.

12. Within Western theatre dance, examples of abstract, formalist dance can be seen in both the ballet and modern dance canon. Examples of pure formalism within these genres are best exemplified in the work of Balanchine and Cunningham (Mackrell, 1991). Although 1980s postmodern dance, the period out of which video dance has emerged, has been characterized as a 'rebirth of content' (Banes, 1987), formalist and abstract devices can still be employed. For instance, the use of repetition, a common strategy within postmodern dance forms, is an example of formal exploration. Therefore, as Chapter 4 is particularly concerned with the influence of postmodern stage dance within the television context, any discussion of formalist devices is relevant to postmodern dance practices in addition to the wider history of formalism within ballet and modern dance traditions.

13. Narrative is paid to be structured in a similar way to language. Stam et al. (1992, p. 9) state that 'the identity of any linguistic sign is determined by the sum total of paradigmatic and syntagmatic relations into which it enters with other linguistic signs in the same language system'. As already suggested, syntagmatic relations are characterized through the sequential arrangement of linguistic (or cinematic) signs. For instance, the sentence 'I like roses' is organized in a particular syntagmatic order; it is a horizontal chain of signs that is meaningful in its arrangement. Meanwhile the paradigmatic structure of language (or cinematic images), relates to a vertical set of units based on similarity or contrast. For example, the word 'roses' in the above sentence could be replaced with an alternative sign, such as tulips or daisies, from the paradigm of flowers, or the word 'like' could be substituted by the oppositional notion of 'hate'. Hence, the paradigmatic choice of linguistic (or cinematic) sign is based on its differential or oppositional status to other related signs.

14. The 'fabula' 'is usually understood as the raw material or basic outline of the story, prior to its artistic organization' (Stam et al., 1992, p. 71); and the 'syuzhet' 'can be understood as the artistic organization, or "deformation," of the causal-chronological order of events' (Stam et al., 1992, p. 71).

15. Once again, it is important to note that the spectator may take up identification positions that resist the dominant or preferred reading.

16. More recently, film theory has addressed other areas of representation, such as homosexuality and ethnicity (Peckham, 1990; Bhabha, 1992; Dyer, 1993), and they too are inscribed with hegemonic discourses.

17. 'A series has the same lead characters in each episode, but each episode has a different story which is concluded. There is "dead time" between the episodes, with no memory from one to the other, and episodes can be screened or repeated in any order ... Serials, on the other hand, have the same characters, but have continuous storylines, normally more than one, that continue from episode to episode' (Fiske, 1989, p. 150).

18. In narrative fictions, although the events need not be relayed in a linear fashion, the reader usually has some understanding of the temporal order of events. For instance, some events may be told as part of a 'flashback' scenario, or some events cut back and forth from one another to suggest that they occur simultaneously. The reader, however, retains some sense of the original linear order of events. In video dance the order of events is far more arbitrary. Although they are obviously edited to the film maker's

specification, the events could ostensibly be played back in any order as they are not subject to a pre-existing linear structure.

19. Within cinema there are instances of the series form such as the *Mad Max* (1979, 1981 and 1985) trilogy and the *Nightmare on Elm Street* (1984, 1985, 1987, 1988 and 1989) collection.

20. Metz insists that the spectator acknowledges that cinema is imaginary and is thus able to accept these fictional and, in many instances, improbable narratives. The spectator is aware that cinema involves a process of perception as the film is brought into existence through machinery. For instance, if the image tilts, the spectator is aware that the camera has rotated, as she or he has not moved her or his head. Hence, it is the spectator's own act of perception and, more specifically, the act of looking, that she or he first identifies with (Metz, 1975; Stam et al., 1992).

21. Although Metz's theory of identification was developed in relation to film, this particular area of theory is equally applicable to television.

22. It has been argued, however, that in the same way that spectators do not always fully engage with television viewing, nor do they always fully engage with theatre, film and visual art (Brunsdon, 1990). It could also be suggested that although some spectators engage only intermittently with the flow of television, other spectators choose to watch specific programmes with the degree of attention that is associated with theatre and film.

23. 'It is a virtual cliché that TV is the most potent means to show the benefits of a product. And that showing is more persuasive than telling or saying' (Evans, 1988, p. 46).

5 Hybrid Sites and Fluid Bodies

1. Scholars from the cultural studies field have also commented on the similarities between television advertisements and music video (Frith, 1988; Kaplan, 1988b; Goodwin, 1993; Falk, 1994).

2. It is important to note, however, that the postmodernist characterization of music video has also been firmly challenged (Goodwin, 1993; Straw, 1993).

3. For a technical explanation of 'pixillation' see Chapter 3, note 9.

4. Alison Murray regularly directs music videos, Lea Anderson choreographed an advertisement for the Ford Ka and Wayne McGregor choreographed and performed in an advertisement for Eurostar.

5. This notion of conservatism is also pertinent to music television. MTV was initially criticized for its censorship policy of excluding music videos by black bands and any references to explicit sex (Kaplan, 1988b).

6. Even when certain values are disrupted in some way that results in a division of opinion, such as changing gender roles, advertising finds ways in which to reconcile these ideological tensions. For instance, a single text may promote a number of readings to appeal to the widest possible audience. Constructions of femininity may include images of assertive, professional females; yet they are slim and attractive, so that the advertisement appeals to both a feminist reader and a conservative reader.

7. On the week ending 19 December 1999 16.01 million people tuned in to *Coronation Street* and 13.53 million people watched a regular episode of *Eastenders* (*The Sunday Times*, 9 January 2000, Culture Section, p. 58, *source*: BARB).

8. See the section titled 'Histories of Dance on Screen' in Chapter 1 for an explanation of *Lifeforms*.

9. Film or video have been employed in Bausch's *Waltzes* (1982), Larrieu's *Waterproof* (1986), Lock's *New Demons* (1987), Forsythe's *Slingerland* (1990) and Decouflé's *Abracadabra* (1999).

10. The obvious exception to this is live performance which integrates digital technologies.

11. Although these are well-known Marxist themes, it is notable that Marx's economic theories are based on an abstract model rather than an empirical reality (McLellan, 1975) and his ideas moved on in later writings (Appelbaum, 1988)

12. This quote is taken from an interview with Emma Gladstone that took place on 17 July 1996.

13. This figure is taken from the 'BBC2/Arts Council Dance for the Camera 1994 Information and Guidelines'. Anecdotal evidence from a number of television producers would suggest that video dance constitutes low-budget television.

14. Television series such as *Star Trek*, *Dr Who* and *Blake Seven*, and films such as *Blade Runner* (1982), *The Terminator* (1984), *RoboCop* (1987) and *Lawnmower Man* (1992) all deal with the relations between technology and society.

15. Once again, performances that integrate an element of interactive digital technology would be an obvious exception.

Bibliography

Allen, D. 'Screening Dance', in S. Jordan and D. Allen (eds), *Parallel Lines: Media Representations of Dance*, London: Libbey, 1993.

Anderson, J. 'Digital Aesthetic', *Animated* (Summer 1996) 12–13.

Appelbaum, R. *Karl Marx* (London: Sage, 1988).

Appignanesi L. (ed.), *Postmodernism: ICA Documents* 4 (London: ICA, 1986).

Armes, R. *Patterns of Realism* (London: Tantivy, 1971).

Balsamo, A. 'Forms of Technological Embodiment: Reading the Body in Contemporary Culture', in M. Featherstone and R. Burrows (eds), *Cyberspace Cyberbodies Cyberpunk: Cultures of Technological Embodiment* (London: Sage, 1995).

Banes, S. *Terpsichore in Sneakers* (Hanover: Wesleyan University Press, 1987).

Barnes, C. 'That's dancing', *Dance & Dancers*, 425 (May 1985) 12–13.

Baudrillard, J. *Symbolic Exchange and Death* (London: Sage, 1993).

Bayston, M. 'Dancers on Television', *Dancing Times*, LXXVII, 921 (June 1987) 707.

Bayston, M. 'Dancers on Television', *Dancing Times*, LXXXII, 982 (July 1992) 950.

Benjamin, W. *Illuminations* (London: Fontana, 1973).

Benton, T. *Modern Art and Modernism: Italian Futurism*, Block VI Unit 14 (Milton Keynes: OUP, 1983).

Berger, J. *Ways of Seeing* (Middlesex: Penguin, 1972).

Bhabha, H. 'The Other Question: the Stereotype and Colonial Discourse' in J. Caughie and A. Kuhn (eds), *The Sexual Subject: a Screen Reader in Sexuality* (London: Routledge, 1992).

Billson, A. 'Flashdance' in J. Pym (ed), *Time Out Film Guide*, 4th edn (London: Penguin, 1995) 248.

Boddington, G. 'Fluid Space, Fluid Presence', *Animated* (Spring 1999) 14–16.

Boyne R. and Rattansi A. (eds) *Postmodernism and Society* (London: Macmillan, 1990).

Bozzini, A. 'They Film as they Dance', *Ballett International*, 14, 1 (January 1991) 37–8, 40.

Brennan, M. 'Mark Baldwin Makes a Pointe', *Dance Theatre Journal*, 13, 1 (Summer 1996) 18–21.

Briginshaw, V. 'Do we Really Know What Post-modern Dance is? Or Copeland's 'Objective Temperament' Revisited', *Dance Theatre Journal*, 6, 2 (Autumn 1988) 12–13, 24.

Briginshaw, V. 'Getting the Glamour on our own Terms', *Dance Theatre Journal*, 12, 3 (Winter 1995/96) 36–9.

Brooks, V. 'Conventions in the Documentary Recording of Dance: Research Need', *Dance Research Journal*, 19, 2 (Winter 1987–88) 15–26.

Brooks, V. 'Dance and Film', *Ballett International* (February 1993) 22–25.

Brown, C. *Inscribing the Body: Feminist Choreographic Practices*, unpublished PhD thesis, University of Surrey (1994).

Brown, I. 'Shown up by the Fish Dish', *Daily Telegraph*, (9 January 1994) Review 10.

Brunsdon, C. 'Television: Aesthetics and Audiences', in P. Mellencamp (ed.), *Logics of Television: Essays in Cultural Criticism* (Bloomington and Indianapolis: Indiana University Press, 1990).

Buckland, T. 'Dance & Music Video' in S. Jordan and D. Allen (eds), *Parallel Lines: Media Representations of Dance* (London: Libbey, 1993).

Burnside, F. 'Moving Pictures on a Black Marble Frame', *Dance Theatre Journal*, 11, 3 (Autumn 1994) 14–17.

Burt, R. *The Male Dancer: Bodies, Spectacle, Sexualities* (London: Routledge, 1995).

Caughie, J. and Kuhn, A. (eds) *The Sexual Subject: A Screen Reader in Sexuality* (London: Routledge, 1992).

Chaurand, J. 'New Art Form with Infinite Possibilities', *Ballett International* (April 1993) 20–1.

Clark, A., Hodson M. and Neiman, C. *The Legend of Maya Deren: A Documentary Biography and Collected Works*, Vol. I, P. One: Signatures (1917–42) (New York: Anthology Film Archives, 1984).

Clark, A., Hodson M. and Neiman C. *The Legend of Maya Deren: A Documentary Biography and Collected Works*, Vol. I, P. Two: Chambers (1942–47) (New York: Anthology Film Archives, 1988).

Clark, N. 'Rear-view Mirrorshades: the Recursive Generation of the Cyberbody', in M. Featherstone and R. Burrows (eds), *Cyberspace Cyberbodies Cyberpunk: Cultures of Technological Embodiment* (London: Sage, 1995).

Collins, J. *Uncommon Cultures: Popular Culture and Post-modernism* (London: Routledge, 1989).

Comolli J. L. and Narboni, P. 'Cinema/Ideology/Criticism', *Screen*, 12, 1 (Spring 1971) 27–36.

Comolli, J. L. and Narboni, P. 'Machines of the Visible', in T. Druckrey (ed), *Electronic Culture: Technology and Visual Representation* (London: Aperture, 1996).

Connor, S. *Postmodernist Culture* (Oxford: Blackwell, 1989).

Constanti, S. 'Exit no Exit: a Contemporary Asian View of the Adam and Eve Myth', *Dance Theatre Journal*, 6, 1 (Summer 1988) 38–9.

Cook, G. *The Discourse of Advertising* (London: Routledge, 1992).

Cook, P. (ed) *The Cinema Book* (London: BFI, 1985).

Copeland, R. 'The Objective Temperament: Post-modern Dance and the Rediscovery of Ballet', *Dance Theatre Journal*, 4, 3 (Autumn, 1986) 6–11.

Copeland, R. 'Abstraction and Hysteria: The Place of the Body in American Non-Literary Theatre', in C. Jones and J. Lansdale (eds), *Border Tensions: Dance and Discourse* (Guildford: University of Surrey, 1995).

Craine, D. 'Dance for the Camera II', *The Times*, (4 February 1995) Vision 6.

Crary, J. and Kwinter, S. (eds) *Incorporations* (New York: Zone, 1992).

Creed, B. *The Monstrous Feminine: Film, Feminism, Psychoanalysis* (London: Routledge, 1993).

Daly, A. (ed.) 'What has Become of Postmodern Dance?', *The Drama Review*, 36, 1 (Spring 1992) 48–69.

Delamater, J. *Dance in the Hollywood Musical* (Michigan: UMI, 1981).

Docherty, T. (ed.) *Postmodernism: a Reader* (Hemel Hempstead: Harvester Wheatsheaf, 1993).

Dodds, S. 'Lea Anderson and the Age of Spectacle', *Dance Theatre Journal*, 12, 3 (Winter 1995/96) 31–3.

Dodds, S. *Video Dance: Hybrid Sites and Fluid Bodies*, unpublished PhD thesis, University of Surrey (1997).

Dodds, S. 'Promoting Pop', *Dance Theatre Journal*, 15, 3 (1999) 8–11.

Druckrey, T. (ed), *Electronic Culture: Technology and Visual Representation* (London: Aperture, 1996)

Dyer, G. *Advertising as Communication* (London: Routledge, 1988).

Dyer, R. 'Stars as Signs', in T. Bennett (ed), *Popular Television and Film* (London: BFI, 1981).

Dyer, R. *The Matter of Images: Essays on Representations* (London: Routledge, 1993).

Eglash, R. 'African Influences in Cybernetics', in C. Hables Gray (ed), *The Cyborg Handbook* (London: Routledge, 1995).

Eisele, H. 'Recorded and Yet Moving Pictures, Dance Culture on Television', *Ballett International*, 13, 9 (September 1990) 15–16.

Evans, R. *Production & Creativity in Advertising* (London: Pitman, 1988).

Falk, P. *The Consuming Body* (London: Sage, 1994).

Featherstone, M. 'The Body in Consumer Culture', in M. Featherstone, M. Hepworth and B. Turner (eds), *The Body: Social Process and Cultural Theory* (London: Sage, 1991).

Featherstone, M. and Burrows, R. (eds) *Cyberspace Cyberbodies Cyberpunk: Cultures of Technological Embodiment* (London: Sage, 1995).

Feuer, J. *The Hollywood Musical* (London: Macmillan, 1982).

Fischer, L. 'The Image of Woman as Image: The Optical Politics of *Dames*' in R. Altman (ed), *Genre: the Musical* (London: Routledge, 1986).

Fiske, J. *Television Culture* (London: Routledge, 1989).

Fiske, J. and Hartley, J. *Reading Television* (London: Methuen, 1978).

Fitting, P. 'The Lessons of Cyberpunk', in C. Penley and A. Ross (eds) *Technoculture* (Oxford: University of Minnesota Press, 1991).

Foster, S. L. 'Dancing Bodies', in J. Crary and S. Kwinter (eds), *Incorporations* (New York: Zone, 1992a).

Foster, S. L. 'Postbody, Multibodies?' in A. Daly (ed), 'What has Become of Postmodern Dance?', *The Drama Review*, 36, 1 (Spring 1992b) 67–69.

Friedberg, A. 'A Denial of Difference: Theories of Cinematic Identification', in E. A. Kaplan (ed), *Psychoanalysis and Cinema* (London: Routledge, 1990).

Frith, S. *Music for Pleasure: Essays in the Sociology of Pop* (Cambridge: Polity, 1988).

Frith, S. 'Youth/Music/Television', in S. Frith, A. Goodwin and L. Grossberg (eds), *Sound and Vision: the Music Video Reader* (London: Routledge, 1993).

Gasché, R. 'Objective Diversions: on Some Kantian Themes in Benjamin's 'The Work of Art in the Age of Mechanical Reproduction', in A. Benjamin and P. Osborne (eds), *Walter Benjamin's Philosophy: Destruction and Experience* (London: Routledge, 1994).

Gibson, W. *Neuromancer* (London: Harper Collins, 1984).

Goldman, R. *Reading Ads Socially* (London: Routledge, 1992).

González, J. 'Envisioning Cyborg Bodies: Notes from Current Research', in C. Hables Gray (ed.), *The Cyborg Handbook* (London: Routledge, 1995).

Goodwin, A. *Dancing in the Distraction Factory: Music Television and Popular Culture* (London: Routledge, 1993).

Gow, G. 'Flashdance and One From the Heart', *Dancing Times*, LXXII, 876 (September 1983) 940.

Grant, S. 'Scent of a Woman', in J. Pym (ed), *Time Out Film Guide*, 4th edn (London: Penguin, 1995) 643.

Grau, A. 'On the Notion of Bodily Intelligence: Cognition, Corporeality and Dance' in C. Jones and J. Lansdale (eds), *Border Tensions: Dance and Discourse* (Guildford: University of Surrey, 1995).

Grossberg, L. 'The Media Economy of Rock Culture: Cinema, Postmodernity and Authenticity', in S. Frith, A. Goodwin and L. Grossberg (eds), *Sound and Vision: the Music Video Reader* (London: Routledge, 1993).

Grossman, P. 'Talking with Merce Cunningham About Video', *Dance Scope*, 13, 2/3 (Winter/Spring 1979) 56–68.

Hables Gray, C. (ed) *The Cyborg Handbook* (London: Routledge, 1995).

Hanna, J. *Dance and Stress: Resistance, Reduction and Euphoria* (New York: AMS, 1988a).

Hanna, J. *Dance, Sex and Gender: Signs of Identity, Dominance, Defiance and Desire* (London: University of Chicago Press, 1988b).

Hansen, S. 'Real-time events', *Dance Theatre Journal*, 14, 3 (1998) 13–15.

Haraway, D. *Simians, Cyborgs and Women: The Reinvention of Nature* (London: Free Association Books, 1991).

Harris, D. *A Society of Signs?* (London: Routledge, 1996).

Hartouni, V. 'Containing Women: Reproductive Discourses in the 1980s', in C. Penley and A. Ross (eds), *Technoculture* (Oxford: University of Minnesota Press, 1991).

Hess, D. 'On Low tech Cyborgs', in C. Hables Gray (ed), *The Cyborg Handbook* (London: Routledge, 1995).

Holland, S. 'Descartes Goes to Hollywood: Mind, Body and Gender in Contemporary Cyborg Cinema', in M. Featherstone and R. Burrows (eds), *Cyberspace Cyberbodies Cyberpunk: Cultures of Technological Embodiment* (London: Sage, 1995).

Hungate, C. (ed) *Totally Wired* (London: ICA, 1996).

Hutcheon, L. *The Politics of Postmodernism* (London: Routledge, 1989).

Jameson, F. *Postmodernism or the Cultural Logic of Late Capitalism* (London: Verso, 1991).

Johnston, C. 'Femininity and the Masquerade: *Anne of the Indies*' in E. A. Kaplan (ed), *Psychoanalysis and Cinema* (London: Routledge, 1990).

Jones, C. 'Computer Conundrums', *Animated* (Summer 1996) 10–11.

Jordan, S. 'Dance Screen 1992', *Dancing Times*, LXXXII, 984 (September 1992) 1154–5.

Jordan, S. and Allen, D. (eds) *Parallel Lines: Media Representations of Dance* (London: Libbey, 1993).

JN, 'Preview: Dance for the Camera', *Guardian* (4 February 1995) The Guide 70.

Kaplan, E. A. (ed) *Postmodernism and its Discontents* (London: Verso, 1988a).

Kaplan, E. A. *Rocking Around the Clock: Music Television, Postmodernism and Consumer Culture* (London: Routledge, 1988b).

Kaplan, E. A. (ed) *Psychoanalysis and Cinema* (London: Routledge, 1990).

Keidan, L. 'Motion Capture', *Animated* (Winter 1998–99) 35, 37–8.

Kellner, D. *Media Culture* (London: Routledge, 1995).

Koch, G. 'Cosmos in Film: on the Concept of Space in Walter Benjamin's 'Work of Art' Essay', in A. Benjamin and P. Osborne (eds), *Walter Benjamin's Philosophy: Destruction and Experience* (London: Routledge, 1994).

Koegler, H. *The Concise Oxford Dictionary of Ballet*, 2nd edn (Oxford: Oxford University Press, 1982).

Kozel, S. 'Spacemaking: Experiences of a Virtual Body', *Dance Theatre Journal*, 11, 3 (Autumn 1994) 12–13, 31, 46–7.

Kozel, S. 'The Virtual World: New Frontiers for Dance and Philosophy', in C. Jones and J. Lansdale (eds), *Border Tensions: Dance and Discourse* (Guildford: University of Surrey, 1995).

Kuhn, A. *The Power of the Image: Essays on Representation and Sexuality* (London: Routledge, 1985).

Landsberg, A. 'Prosthetic Memory: *Total Recall* and *Blade Runner*,' in M. Featherstone and R. Burrows (eds), *Cyberspace Cyberbodies Cyberpunk: Cultures of Technological Embodiment* (London: Sage, 1995).

de Lauretis, T. *Alice Doesn't: Feminism Semiotics Cinema* (London: Macmillan, 1984).

Lockyer, B. 'Dance and Video: Random Thoughts', *Dance Theatre Journal*, 1, 4 (Autumn 1983) 13–16.

Lockyer, B. 'Stage Dance on Television', in S. Jordan and D. Allen (eds), *Parallel Lines: Media Representations of Dance* (London: Libbey, 1993).

Lupton, D. 'The Embodied Computer/user' in M. Featherstone and R. Burrows (eds), *Cyberspace Cyberbodies Cyberpunk: Cultures of Technological Embodiment* (London: Sage, 1995).

Macauley, W. 'From Cognitive Psychologies to Mythologies: Advancing Cyborg Textualities for a Narrative of Resistance', in C. Hables Gray (ed), *The Cyborg Handbook* (London: Routledge, 1995).

Mackrell, J. 'Fugue', *Dance Theatre Journal*, 6, 1 (Summer 1988) 28–9.

Mackrell, J. 'Post-modern Dance in Britain: an Historical Essay', *Dance Research*, 9, 1 (Spring 1991) 40–57.

Mackrell, J. 'Making Television Dance', *Dance Theatre Journal*, 1, 4 (Autumn 1993) 28–9.

Mackrell, J. 'As Seen on Television', *Independent* (8 January 1994) Weekend Arts 44.

Mackrell, J. *Reading Dancing* (London: Michael Joseph, 1997).

Maletic, V. 'Videodance–Technology–Attitude Shift', *Dance Research Journal*, 19, 2 (Winter 1987–88) 3–7.

Manning, S. 'Modernist Dogma and Post-modern Rhetoric', *The Drama Review*, 32, 4 (Winter 1988) 32–9.

de Marigny, C. 'Dance for and on Television', *Dance Theatre Journal*, 6, 1 (Summer 1988) 2–3.

de Marigny, C. 'Video Danse Grand Prix', *Dance Theatre Journal*, 8, 1 (Summer 1990) 30–1.

de Marigny, C. 'Viewing Figures', *Dance Theatre Journal*, 8, 4 (Spring 1991) 34–5.

de Marigny, C. 'Progressive Programming', in S. Jordan and D. Allen (eds), *Parallel Lines: Media Representations of Dance* (London: Libbey, 1993).

Martin, G. 'Readers, Viewers and Texts' in *Popular Culture: Form and Meaning*, 4, 13 (Milton Keynes: OUP, 1981).

McCabe, C. 'Realism and the Cinema: Notes on Some Brechtian Theses', in T. Bennett (ed), *Popular Television and Film* (London: BFI, 1981).

McCarron, K. 'Corpses, Animals, Machines and Mannequins: the Body and Cyberpunk', in M. Featherstone and R. Burrows (eds), *Cyberspace Cyberbodies Cyberpunk: Cultures of Technological Embodiment* (London: Sage, 1995).

McLellan, D. *Marx* (London: Fontana, 1975).

McLuhan, M. *Understanding Media: The Extensions of Man* (London: Routledge & Kegan Paul, 1964).

McRobbie, A. 'Fame, Flashdance and Fantasies of Achievement', in J. Gaines and C. Herzog (eds), *Fabrications: Costume and the Female Body* (London: Routledge, 1990).

Meisner, N. 'Contact or Substitute?', *Dance and Dancers* (March 1984) 24–6.

Meisner, N. 'A Particular Affinity', *Dance and Dancers* (October 1991) 16–18.

Meisner, N. 'Aspiration and Aberration', *Dance and Dancers*, 515 (November/ December 1993) 12–14.

Mercer, C. 'Pleasure', in *Popular Culture: Form and Meaning*, 14, 17 (Milton Keynes: OUP, 1981).

Merrett, S. 'Video Dance: IMZ Dance Screen in Frankfurt', *Dancing Times*, LXXXI, 963 (December 1990) 256–57.

Merrett, S. 'Spotlight on Ross MacGibbon', *Dancing Times*, LXXXII, 980 (May 1992) 746–47

Metz, M. *Film Language: a Semiotics of the Cinema* (New York: OUP, 1974).

Metz, M. 'The Imaginary Signifier', *Screen*, 16, 2 (Summer 1975) 14–76.

Monaco, J. *How to Read a Film* (Oxford: OUP, 1981).

Morley, D. 'Interpreting Television', in *Popular Culture: Form and Meaning*, 3, 11 and 12 (Milton Keynes: OUP, 1981).

Morley, D. 'Television: Not so Much a Visual Medium, More a Visible Object', in C. Jenks (ed), *Visual Culture* (London: Routledge, 1995).

Mueller, J. 'Watching an American Screen Original: Astaire-style Film', *Dance Magazine*, LVIII, 5 (May 1984) 131–5.

Müller, H. 'Expression in Dance', *Dance Theatre Journal* 2, 1 (January, 1984) 10–15.

Mulvey, L. *Visual and Other Pleasures* (London: Macmillan, 1989).

Nagrin, D. 'Nine Points on Making your own Dance Video or How to Really Dance Forever', *Dance Theatre Journal*, 6, 1 (Summer 1988) 33–6.

Nava, M. 'Framing Advertising: Cultural Analysis and the Incrimination of Visual Texts', in M. Nava, A. Blake, I. MacRury and B. Richards (eds), *Buy this Book: Studies in Advertising and Consumption* (London: Routledge, 1997).

Nears, C. 'Bridging a Distance', *Dance Research*, 5, 2 (Autumn 1987) 43–59.

Newman, B. 'Speaking of Dance: Geoff Dunlop', *Dancing Times*, LXXV, 986 (May 1985) 685–6.

Norris, C. *Derrida* (London: Fontana, 1987).

O' Donohoe, S. 'Leaky Boundaries: Intertextuality and Young Adult Experiences of Advertising', in M. Nava, A. Blake, I. MacRury and B. Richards (eds), *Buy this Book: Studies in Advertising and Consumption* (London: Routledge, 1997).

Pascal, J. 'Freefall and Gaby Agis', *Dance Theatre Journal*, 6, 1 (Summer 1988) 42–3.

Parry, J. 'Strange Fish to Goggle at', *Observer* (9 January 1994) Review 12.

Parry, J. 'Site-Seeing Tour', *Dance Now*, 7, 4 (Winter 1998–99) 9–14.

Peckham, L. 'Not Speaking with Language/Speaking with no Language', in E. A. Kaplan (ed), *Psychoanalysis and Cinema* (London: Routledge, 1990).

Penley, C. and Ross A. (eds), *Technoculture* (Oxford: University of Minnesota Press, 1991).

Penman, R. *A Catalogue of Ballet and Contemporary Dance in the Television Film and Videotape Library, 1937–1984* (London: BBC, 1986).

Penman, R. 'Making History: the BBC Television Film and Videotape Library 1937–1984', *Dance Research*, 5, 2 (Autumn 1987) 61–8.

Penman, R. 'Ballet and Contemporary Dance on British Television', in S. Jordan and D. Allen (eds), *Parallel Lines: Media Representations of Dance* (London: Libbey, 1993).

Penman, R. 'Dance on Television', *Dancing Times* (September 1994) 1172–3.

Penman, R. 'Dance on Television', *Dancing Times* (March 1995) 571 and 573.

Penman, R. '2 Dance or Not 2 dance? That is the Question–BBC2', *Dancing Times* (September 1996) 1121–5.

Percival, J. 'Merce Cunningham's Evergreen Commitment', *The Times* (27 June 1980) 11

Pritchard, J. 'Movement on the Silent Screen', *Dance Theatre Journal*, 12, 3 (Winter 1995/96) 26–30.

Rawdon Wilson, R. 'Cyber (Body)parts: Prosthetic Consciousness', in M. Featherstone and R. Burrows (eds), *Cyberspace Cyberbodies Cyberpunk: Cultures of Technological Embodiment* (London: Sage, 1995).

Rheingold, H. *Virtual Reality* (London: Secker & Warburg, 1991).

Robertson, A. 'Smart Moves', *Time Out*, 1276 (1–8 February 1995) 157.

Rosiny, C. 'Pina Bausch "The Lament of the Empress"', *Ballett International*, 13 (June/July 1990a) 74.

Rosiny, C. 'Dance as Seen from the Perspective of the Camera', *Ballett International*, 13, 9 (September 1990b) 8, 10, 12.

Rosiny, C. 'Dance Films and Video Dance', *Ballett International* (August/ September 1994) 82–3.

Ross, A. 'Hacking Away at the Counterculture', in C. Penley and A. Ross (eds), *Technoculture* (Oxford: University of Minnesota Press, 1991).

Rowson Davis, J. 'Ballet on British Television, 1933–1939', *Dance Chronicle*, 5, 3 (1982–83) 245–304.

Rubidge, S. 'New Criteria New Alternatives', *Dance Theatre Journal*, 2, 4 (Winter 1984) 36–8.

Rubidge, S. 'Dancelines 2', *Dance Theatre Journal*, 6, 1 (Summer, 1988a) 6–9.

Rubidge, S. 'Strong Language: from Stage to Screen', *Dance Theatre Journal*, 6, 1 (Summer 1988b) 11–13, 17.

Rubidge, S. 'Steps in Time – A television Director's Perspective', *Dance Theatre Journal*, 6, 1 (Summer 1988c), 15–17.

Rubidge, S. 'Recent Dance Made for Television', in S. Jordan and D. Allen (eds), *Parallel Lines: Media Representations of Dance* (London: Libbey, 1993).

Rubidge, S. 'Defining Digital Dance', *Dance Theatre Journal*, 14, 4 (1999) 41–5.

Sacks, A. 'Dance that Passes the Screen Test', *Independent on Sunday* (9 January 1994) Arts 24.

Sarup, M. *Post-structuralism and Postmodernism* (Hemel Hempstead: Harvester Wheatsheaf, 1988).

Satin, L. 'Movement and the Body in Maya Deren's "Meshes of the Afternoon"', *Women and Performance*, 6, 2, Issue 12 (1991) 41–9.

Schmidt, J. 'Exploitation or Symbiosis? On the Contradictions Between Dance and Video', *Ballett International*, 14, 1 (January 1991) 97–9.

Shilling, C. *The Body and Social Theory* (London: Sage, 1993).

Siegel, M. 'Is it Still Postmodernism? Do we Care?' in A. Daly (ed), 'What has Become of Postmodern Dance?', *The Drama Review*, 36, 1 (Spring 1992) 49–51.

Silverstone, R. *The Message of Television: Myth and Narrative in Contemporary Culture* (London: Heinemann, 1981).

Slater, D. 'Consumer Culture and the Politics of Need', in M. Nava, A. Blake, I. MacRury and B. Richards (eds), *Buy this Book: Studies in Advertising and Consumption* (London: Routledge, 1997).

Stacey, J. 'Desperately Seeking Difference', in J. Caughie and A. Kuhn (eds), *The Sexual Subject: a Screen Reader in Sexuality* (London: Routledge, 1992)

Stam, R., Burgoyne, R. and Flitterman-Lewis, S. *New Vocabularies in Film Semiotics: Structuralism, Post-structuralism and Beyond* (London: Routledge, 1992).

Straw, W. 'Popular Music and Postmodernism in the 1980s', in S. Frith, A. Goodwin and L. Grossberg (eds), *Sound and Vision: the Music Video Reader* (London: Routledge, 1993).

Stoop, N. 'Footloose: On Location with Herbert Ross', *Dance Magazine*, 58, 3 (March 1984) 58–60.

Tegeder, U. 'Dance and Video', *Ballett International*, 8, 2 (February 1985) 24–6.

Tidsall, C. and Bozzolla, A. *Futurism* (London: Thames and Hudson, 1977).

Tomas, D. 'Feedback and Cybernetics: Reimaging the Body in the Age of Cybernetics', in M. Featherstone and R. Burrows (eds), *Cyberspace Cyberbodies Cyberpunk: Cultures of Technological Embodiment* (London: Sage, 1995).

Turner, G. *Film as a Social Practice* (London: Routledge, 1993).

Wernick, A. *Promotional Culture: Advertising, Ideology and Symbolic Expression* (London: Sage, 1991).

Willeman, P. 'On Realism in the Cinema', *Screen*, 13, 1 (Spring 1972) 37–45.

Williams, C. *Realism and the Cinema* (London: Routledge & Kegan Paul, 1980).

Williamson, J. *Decoding Advertisements: Ideology and Meaning in Advertising* (London: Marion Boyars, 1978).

Wollen, W. *Signs and Meaning in the Cinema* (London: Secker & Warburg, 1969).

Wyman, M. 'Computer Program [sic] Aids Dance Makers', *Dance Magazine*, LXV, 3 (March 1991) 12–13.

Wyver, J. 'Television and Postmodernism', in L. Appignanesi (ed.), *Postmodernism: ICA Documents 4* (London: ICA, 1986).

Index